Depressive Illness

Advances in Biological Psychiatry

Vol. 7

Series Editors
J. Mendlewicz, Brussels, and *H.M. van Praag,* Utrecht

S. Karger · Basel · München · Paris · London · New York · Sydney

Symposium on Depressive Illness, Amsterdam, September 11–12, 1980

Depressive Illness

Biological and Psychopharmacological Issues

Volume Editors
J. Mendlewicz, Brussels, *A. Coppen,* Epsom, and *H.M. van Praag,* Utrecht

45 figures and 63 tables, 1981

S. Karger · Basel · München · Paris · London · New York · Sydney

Advances in Biological Psychiatry

National Library of Medicine, Cataloging in Publication
Depressive illness, biological and psychopharmacological issues
Volume editors, J. Mendlewicz and H.M. van Praag. – Basel, New York, Karger, 1981.
(Advances in biological psychiatry, v. 7)
Proceedings of the first joint meeting between the Interdisciplinary Society of Biological Psychiatry and the World Psychiatric Association.
Wl Ad44 v. 7 / [WM 171 D4247]
ISBN 3–8055–2482–X

Drug Dosage

The authors and the publisher have exerted every effort to ensure that drug selection and dosage set forth in this text are in accord with current recommendations and practice at the time of publication. However, in view of ongoing research, changes in government regulations, and the constant flow of information relating to drug therapy and drug reactions, the reader is urged to check the package insert for each drug for any change in indications and dosage and for added warnings and precautions. This is particularly important when the recommended agent is a new and/or infrequently employed drug.

Printed in Switzerland by Thür AG, Offsetdruck, Pratteln
ISBN 3–8055–2482–X

Contents

Introduction

This is the first joint meeting between the Interdisciplinary Society of Biological Psychiatry and the World Psychiatric Association (WPA).

During the last 30 years, we have witnessed significant progress in the management, treatment and understanding of depressive illness. Still, basic questions such as the mechanisms of action of antidepressants, ECT and lithium remain unclear, and although most researchers are today convinced that many types of depression have a biological substrate, unequivoqual evidence of one or more specific biological abnormality in depression is still lacking. One more puzzling issue, we feel, is the problem of classification of depressive subgroups (often raised in depressive research) and the necessity to support clinical classification with specific and reliable biological parameters. These issues will be discussed in this publication.

J. Mendlewicz, Brussels

Adv. biol. Psychiat., vol. 7, pp. 1–13 (Karger, Basel 1981)

Standardization of the Dexamethasone Suppression Test for the Diagnosis of Melancholia[1]

Bernard J. Carroll, Michael Feinberg, John F. Greden, Janet Tarika, A. Ariav Albala, Roger F. Haskett, Norman McI. James, Ziad Kronfol, Naomi Lohr, Meir Steiner, Jean Paul de Vigne, Elizabeth Young

Clinical Studies Unit and Mental Health Research Institute, Department of Psychiatry, University of Michigan, Ann Arbor, Mich., USA

Depression is a very common problem in medical practice and about 70% of prescriptions for tricyclic antidepressant drugs are written by physicians in family practice or internal medicine [1]. Despite the widespread use of these drugs, depressed patients are frequently not diagnosed or treated adequately [1, 2]. The diagnosis of patients with psychiatric disorders has traditionally been based on the clinical history and on behavioral observations. Objective laboratory diagnostic tests for the so-called functional psychiatric disorders have not been available. The reliability and validity of psychiatric diagnoses have been studied closely in recent years. Standardized diagnostic criteria have been proposed [3, 4], and validation has been attempted by investigations of genetic associations, response to treatment, or long-term clinical course [5, 6]. Despite these efforts, there is a lack of general consensus on many issues among psychiatrists. The differential diagnosis of patients with depression has been a particularly controversial area [7, 8], and a reliable laboratory test to assist in making the diagnosis of melancholia would be useful.

We have previously reported a test of neuroendocrine function that is highly specific for melancholia [9–11]. Inpatients with melancholia were found to have an abnormal early escape of plasma cortisol concentrations during a 24-hour overnight dexamethasone suppression test (DST). Other investigators have confirmed this finding among inpatients [12–18], and we have repeated our observation in a study of outpatients [8]. The DST is now being used widely for the diagnosis of melancholia [19], but differing procedures have been adopted in various centers, and the test needs to be standardized for general use. Differing doses of dexamethasone, different times of blood sampling, and different criteria

[1] This manuscript is a condensed version of an article that appeared in Archives of General Psychiatry *38:* 15–22, 1981.

for judging a DST response as abnormal have been used empirically in the various studies of the test reported to date [10–19].

We report here the results of a 6-year study of the test, with data that provide a firm basis for its standardization. Aspects of the test procedure and its diagnostic performance that we have studied include: dose of dexamethasone; number and timing of blood samples; the criterion value of plasma cortisol concentration for judging a test result as abnormal; the possible interference caused by recent use of psychotropic drugs; the effects of age, sex and severity of depressive symptoms; the relative utility in outpatient and inpatient settings; the results obtained in normal subjects; and the sensitivity, specificity and diagnostic confidence obtained in a comparison of melancholic patients with patients suffering from other psychiatric disorders.

Methods

Diagnostic Procedure

Patients were evaluated in the Clinical Studies Unit of the Department of Psychiatry, University of Michigan Medical Center. Each patient was seen by a psychiatrist for a standard clinical assessment, which usually included an interview with a close relative. Other sources of information used were the patient's previous psychiatric history, family history and past hospital records. Patients seen after July 1977 (73% of all cases reported here) also completed a structured psychiatric interview, the Schedule for Affective Disorders and Schizophrenia (SADS) [20], which was administered by a trained social worker, research assistant or psychiatrist. The primary psychiatrist and the SADS interviewer then met for a diagnostic conference where items of information were cross-checked before consensus diagnoses were reached. The clinical features we use for the diagnosis of endogenous and nonendogenous depression have been discussed in detail in a previous publication [8]. In accordance with the newly revised Diagnostic and Statistical Manual of the American Psychiatric Association (DSM III) [21], we will use the term melancholia, rather than endogenous depression. Our clinical diagnoses of melancholia were supported by the RDC [4] in over 95% of cases [8].

We studied a total of 368 patients, 180 inpatients and 188 outpatients. There were 215 with a diagnosis of melancholia. For a comparison group we studied 100 patients with nonendogenous depression and 53 patients with miscellaneous other psychiatric disorders (table I). Thus, the prevalence of melancholia in the total group of patients studied was 58%. For an additional comparison we studied 70 euthymic subjects who had no current psychiatric disorder.

The severity of depressive symptoms was assessed in patients with melancholia and with nonendogenous depression by the Hamilton rating scale [22] and the Carroll rating scale [23], which is a self-rating adaptation of the Hamilton scale.

Exclusion Criteria

Technical factors which can interfere with the DST have been discussed at length [10, 24]. They are severe weight loss, anorexia nervosa, pregnancy, and physical illness such as acute withdrawal from alcohol, uncontrolled cardiac failure or hypertension, fever, dehydration, liver disease, unstable diabetes mellitus, endocrine disorders, and organic brain disease. Patients were also excluded if they had recently been receiving any of the following drugs: barbiturates, anticonvulsants, glucocorticoids, spironolactone, reserpine, narcotics, alpha-

Table I. Details of 4 series of dexamethasone suppression tests

Clinical status	Inpatients		Outpatients		Total
Dexamethasone dose	1 mg	2 mg	1 mg	2 mg	
Melancholia	54	59	49	53	215
Other diagnosis	22	45	58	28	153
Total	180		188		368

methylDOPA, cyproheptadine, or high dosage estrogens. Patients were excluded if they had a drug-induced psychosis, or if they had received electroconvulsive therapy within the previous 3 days. 9% of all outpatients and 20% of all inpatients were excluded from the study for one or more of these reasons. 60% of the technical exclusions were on account of the drugs listed above.

Psychotropic Drugs

Previous use of psychotropic drugs (antidepressants, neuroleptics, lithium, benzodiazepines, chloral hydrate) was documented for all patients. They were kept free of all drugs for as long as was clinically possible. Over 50% of all patients were drug-free for at least 7 days before the DST was performed. The pattern of abnormal DST results was analyzed to determine if there was a relationship with psychotropic drug intake.

DST Procedure

The overnight DST procedure [25, 26] was used. Patients received dexamethasone tablets at 23.30 h. In order to standardize the test we compared two doses of dexamethasone in outpatients and inpatients: 183 patients received 1 mg, while 185 received 2 mg (table I). Blood samples for plasma cortisol determination were obtained from the 180 inpatients at 08.00 h, 16.00 h and 23.00 h on the following day. For practical reasons we obtained only the 16.00 h blood sample from the 188 outpatients, who followed their usual schedule of work or activity except for a visit to the hospital at 16.00 h.

Plasma cortisol was measured by the competitive protein binding method of *Murphy* [27], with minor modifications [10]. The between-assay coefficient of variation was 8%. The samples were assayed without knowledge of the clinical diagnoses, and the results were not released to the clinicians until the diagnostic evaluations were completed.

Statistics

The data were entered on coding forms, stored in a computer file, and analyzed using the Michigan Interactive Data Analysis System (MIDAS). The diagnostic performance of the DST was evaluated with the calculations of sensitivity, specificity and diagnostic confidence (predictive value), as defined by *Vecchio* [28] and by *Galen and Gambino* [29]. The term sensitivity refers to the true-positive rate, or the proportion of melancholic patients in whom abnormal DST results were found. Specificity refers to the true-negative rate, or the proportion of nonmelancholic subjects in whom normal results were observed. Diagnostic confidence refers to the proportion of abnormal test results that were true-positive for melancholia.

The criterion of a normal postdexamethasone plasma cortisol concentration used in our previous studies was set conservatively at 6 μg/dl. In the present set of 368 tests we have compared the diagnostic performance associated with plasma cortisol criterion values of 3, 4, 5 and 6 μg/dl.

In the inpatient series (180 tests) a positive or abnormal DST result was recorded when any one of the three plasma cortisol concentrations was above the criterion value [9, 10]. The relative diagnostic efficacy of the 08.00 h, 16.00 h and 23.00 h postdexamethasone blood samples was examined in the inpatient series. This analysis also will permit the determination of the relative decrease in diagnostic performance associated with limiting the blood sampling to only the 16.00 h sample in the outpatient series.

Logarithmic transformation of the plasma cortisol concentrations was performed in all parametric analyses of the results [30, 31].

Results

Selection of a Criterion Value

The sensitivity and specificity associated with plasma cortisol criterion concentrations between 3 and 6 μg/dl were determined. For these analyses the results of all 215 tests in patients with melancholia and of all 153 tests in other patients, respectively, were used. The criterion value of 5 μg/dl gave an *overall* test sensitivity of 43% and a specificity of 96%. The higher criterion value of 6 μg/dl gave a lower sensitivity (39%) with no significant improvement in specificity (97%). Lower criterion values (3 and 4 μg/dl) were associated with specificity results below 95%. For the remaining analyses a postdexamethasone plasma cortisol criterion value of 5 μg/dl will be used. Using this criterion the diagnostic confidence for melancholia of an abnormal DST result was 94% (92% when adjusted for a standardized prevalence rate of 50%.

Euthymic Subjects

70 euthymic subjects with no current psychiatric diagnosis were tested with the outpatient procedure. Postdexamethasone plasma cortisol concentrations below 5 μg/dl were found in 67 of these 70 subjects. Thus, the specificity of the test was 96% in this group. This result is identical with the specificity found among the 153 patients with diagnoses other than melancholia. From these data it appears that about 4% of normal persons may have false-positive DST results.

Dose of Dexamethasone

The dose of dexamethasone had a notable influence on the sensitivity of the DST (table II). Among inpatients the sensitivity was 67% with 1 mg, 39% with 2 mg (chi-squared, $p < 0.005$). Among outpatients the corresponding figures were 49 and 17% ($p < 0.001$). By contrast the specificity of the test was not affected significantly by the dose of dexamethasone. Combining both inpatients and outpatients at each dose, the specificity was 96% with 1 mg and 96% with 2 mg. The overall diagnostic confidence of a positive test result was 92%. The low diagnostic confidence of 85% found in the 2-mg inpatients series was mainly

Table II. Diagnostic performance of the DST with 2 doses of dexamethasone among inpatients and outpatients: the plasma cortisol criterion value was 5 μg/dl.

Clinical Status	Inpatients		Outpatients		Total
Dexamethasone dose	1 mg	2 mg	1 mg	2 mg	
Sensitivity, %	67	39	49	17	43
Specificity, %	96	93	97	100	96
Predictive value, % (adjusted for 50% prevalence rate)	94	85	94	100	92

due to the low sensitivity, rather than to a major lowering of the specificity (93%) [29].

These results indicate clearly that the 1-mg dose of dexamethasone is the more effective in the routine diagnostic application of the DST for melancholia. The sensitivity will be greatly improved and there will be no loss of specificity. The sensitivity of 67% obtained among inpatients with the 1-mg dose indicates that the general utility of the test will be greater than appeared from earlier inpatient studies with the 2-mg dose [10, 12], in which a sensitivity in the range of 40% was found.

Time of Blood Sampling

The relative diagnostic efficacy of the three blood samples obtained from inpatients (08.00 h, 16.00 h and 23.00 h) was determined. As stated earlier, we considered a positive DST result to be present when the plasma cortisol concentration of any one of the three samples exceeded the criterion value (5 μg/dl). 24% of all the positive tests were detected by the 08.00 h sample, 78% by the 16.00 h sample, and 71% by the 23.00 h sample. The 16.00 h and 23.00 h samples in combination detected 98% of all the positive tests. These results indicate that, for practical purposes, the 16.00 h and 23.00 h samples will suffice for the diagnostic application of the DST in inpatient psychiatric settings. Blood sampling at both times will be necessary for optimal sensitivity: 21% of all the positive tests were detected only by the 16.00 h sample, and 19% only by the 23.00 h sample.

These results also help to clarify the difference in sensitivity noted earlier between the inpatient series and the outpatient series (table II). The inpatient results were based on 3 blood samples, the outpatient results on only the 16.00 h blood sample. Only 78% of all the positive tests among inpatients were detected by the 16.00 h sample alone. Thus, this apparent difference in sensitivity between melancholic inpatients and outpatients needs to be adjusted for the procedural difference between the two series.

Table III. Frequency of positive DST results among melancholic patients in relation to number of days since last use of psychotropic drugs: combined results for 1- and 2-mg doses of dexamethasone in outpatients and inpatients

Number of Days drug-free	1–3	4–6	7–9	>10
Number of patients	98	11	13	93
Frequency of positive DST, %	47	45	46	36

Comparison of Inpatients and Outpatients

When only the 16.00 h postdexamethasone plasma cortisol concentrations were considered we found no significant difference in sensitivity between inpatients and outpatients. 43% of 113 melancholic inpatients, and 32% of 102 melancholic outpatients, had a 16.00 h plasma cortisol concentration above 5 μg/dl (chi-squared = 2.75, $p < 0.10$). The same result was found when the 1- and 2-mg dose series were examined separately. With the 1-mg dose the sensitivity was 57% for 54 inpatients, 49% for 49 outpatients (chi-squared = 0.73, $p < 0.4$). With the 2-mg dose the sensitivity was 30% for 59 inpatients, 17% for 53 outpatients (chi-squared = 2.8, $p < 0.1$). These results indicate the presence of a trend for abnormal DST results to occur more frequently among inpatients than outpatients with melancholia, but the trend was not significant with the number of patients studied so far. The specificity of the test, based on only the 16.00 h postdexamethasone plasma cortisol concentrations, was 98% for both inpatients and outpatients.

Psychotropic Drugs

We observed no significant effect of a history of recent psychotropic drug intake on the DST results. Since positive test results were found almost exclusively among the patients with melancholia, we have restricted the analysis of psychotropic drug effects to this group alone. The results of this analysis are shown in table III, where it can be seen that there was no change in the frequency of positive test results until the length of time off drugs exceeded 10 days. A comparison of the two extreme groups (1–3 vs. ≥10 days drug-free) revealed no significant difference in the frequency of abnormal DST results (chi-squared = 2.4, $p < 0.2$). In view of the effects of dose of dexamethasone, and the trend for an effect of inpatient versus outpatient status on the sensitivity of the test, we examined the effect of psychotropic drugs separately in each of the four series: inpatient, 1 mg; inpatient, 2 mg; outpatient, 1 mg; outpatient, 2 mg. In each case we compared the frequency of positive DST results found in the two extreme groups (1–3 vs. ≥10 days drug-free). In none of the 4 series was a significant difference obtained. The Fisher exact probability calculated in each case was 0.93, 0.69, 0.27, and 0.44, respectively.

Age and Sex

Within the group of 215 melancholic patients there was no significant relationship between age and DST result. The correlation coefficient between age and 16.00 h plasma cortisol concentration was +0.13. This result indicates that 2% of the variance in the plasma cortisol concentrations was accounted for by the factor of age among the melancholic patients. The corresponding correlation coefficient among the 153 nonmelancholic patients was +0.15 (not significant). The mean age of the melancholic patients with a positive DST result was 49 (SD 18) years, compared to 46 (SD 15) years for those with a negative test result ($p < 0.08$). Both young and elderly patients with melancholia were identified by the test.

There was a similar sex distribution among the melancholic and nonmelancholic groups of patients. 64% of the former group and 61% of the latter group were female. Among the melancholic patients positive DST results were found in 37% of the males, and in 46% of the females (chi-squared = 1.57, $p < 0.3$). Thus, male and female patients with melancholia were detected with equivalent sensitivity by the test.

Severity of Depressive Symptoms

The self-rated severity of depression was recorded with the Carroll rating scale (CRS) by 180 (84%) of the 215 melancholic patients, and 80 of the 100 patients with nonendogenous depression. These subgroups with CRS ratings did not differ from their parent groups with respect to age or frequency of abnormal DST results.

The mean CRS score of the 180 melancholic patients was 25.7 (SD 9.7). A very similar mean CRS score was recorded by the 80 patients with nonendogenous depression (mean 23.7, SD 9.5). The difference was not significant ($p > 0.10$). The median score of normal subjects on the CRS is 3 [21]. These CRS scores indicate that the two groups had a similar experience of general depressive symptoms. Thus, the difference in DST results between the two groups was related to the categorical diagnostic distinction between melancholia and nonendogenous depression. It was not related to a difference in the degree of subjective distress between the two groups. Similar results were found with the psychiatrists' ratings of severity by the Hamilton rating scale.

Discussion

The results of this study confirm the diagnostic value of the DST for melancholia, and give a basis for the standardization of this test in psychiatric settings. Our results allow us to state the optimal dose of dexamethasone, the most efficient number of blood samples, and the plasma cortisol criterion value which yields a diagnostic confidence close to 95%.

Before discussing these results in detail, we wish to emphasize that an abnormal DST result is not *biologically* specific for melancholia, and that the technical exclusion criteria we listed must be understood by physicians who use the test. We excluded 9% of all outpatients and 20% of all inpatients for one or more of these technical reasons. The rationale for these exclusion criteria has been discussed at length in previous publications [10, 24].

Among properly screened psychiatric patients, however, an abnormal DST result is highly specific for melancholia. The addition of our present results to those reported by other investigators [12–18] gives a total of 1,058 psychiatric patients investigated (527 with melancholia or an essentially equivalent diagnosis, 531 with other diagnoses). From the combined results of all these studies a sensitivity of 45%, a specificity of 96% and a diagnostic confidence of 92% can be calculated. There were variations in the DST procedures used in these studies, which would be expected to affect the sensitivity rather than the specificity. The plasma cortisol criterion values varied between 4 and 6 μg/dl; the dose of dexamethasone ranged between 0.5 and 2 mg; some investigators collected blood samples only at 08.00 h, some only at 16.00 h, and others used more frequent blood sampling. Our present results indicate that the plasma cortisol criterion value can be lowered from 6 to 5 μg/dl without a significant loss of specificity, and with some increase of sensitivity. The DST results we obtained among 70 euthymic subjects with no current psychiatric diagnosis were very similar to those found among the 153 nonmelancholic psychiatric patients. We do not know why about 4% of the normal subjects had an abnormal DST result. This background rate of false-positive results in normal subjects, however, gives a useful point of reference for evaluating the rates found in other groups of patients.

The dose of dexamethasone used had a major effect on the sensitivity of the DST, but did not affect the specificity. Among inpatients with melancholia the sensitivity was 67% with 1 mg dexamethasone, and 39% with 2 mg (based on the results of 3 blood samples with the inpatient procedure). This represents a 72% relative gain in sensitivity with the 1 mg dose. Among melancholic outpatients (based on only 1 blood sample) the sensitivity was 49% with 1 mg, and 17% with 2 mg. This represents a 188% relative gain in sensitivity with the 1 mg dose. In view of the identical specificity overall (96% with both doses), we recommend that the 1-mg dose of dexamethasone be used in the routine diagnostic application of the DST for melancholia. Doses of dexamethasone lower than 1 mg have been tested in only one study to date. Using 0.5 mg dexamethasone *Nuller and Ostroumova* [18] obtained a sensitivity of 69% and a specificity of 91%, with a diagnostic confidence (adjusted for a 50% prevalence rate) of 88%. These results were found with a single 08.00 h postdexamethasone blood sample, and are not better than our present results with 1 mg (67% sensitivity, 96% specificity, 92% adjusted diagnostic confidence).

The relative diagnostic efficacy of the 3 blood samples obtained from inpatients was consistent with the results obtained in previous, smaller studies [10, 12]. For practical purposes, we can eliminate the 08.00 h blood sample, since 98% of the positive tests were detected by the 16.00 h and 23.00 h samples in combination. This result will improve the cost-effectiveness of the test, since it will allow the laboratory work to be reduced by one-third.

This analysis also revaled that we obtained a 22% loss of sensitivity by limiting the outpatient DST procedure to only the 16.00 h blood sample. This conclusion follows from the finding that only 78% of the positive inpatient tests were detected by the 16.00 h sample alone. The test procedùre we used with outpatients was a compromise for practical reasons. If a second blood sample could be obtained later in the day (about 21.00–23.00 h) we would expect to find an increased sensitivity among outpatients.

After adjusting the sensitivity results for this procedural difference (3 blood samples compared to 1 blood sample), we found no significant difference between inpatients and outpatients with melancholia. There was a trend for the inpatients to have a higher frequency of abnormal DST results. This trend was probably related to a general severity of illness factor [12, 32] which is not well reflected by the standard rating scales.

The frequency of positive DST results among patients with melancholia was not significantly affected by a history of recent use of psychotropic drugs. This finding is in agreement with earlier reports [10, 12, 13] and is reassuring with respect to the general utility of the test. Most patients will be able to tolerate a brief period (2–3 days) of withdrawal from psychotropic drugs in preparation for the test. There are theoretical reasons for expecting that high doses of benzodiazepines could cause spuriously negative test results [18, 33]. Our present results and those of others [12, 13] indicate that a recent history (1–3 days) of use of lithium, antidepressants, neuroleptics, chloral hydrate or low-dose benzodiazepines will not materially affect the DST results.

The age and sex results indicated that these factors did not account for the frequency of abnormal DST results among melancholic patients. Both young and elderly patients, of either sex, will be identified by the test. There was also no meaningful difference in severity of depressive symptoms between melancholic patients and those with nonendogenous depression. Thus, the specificity of the DST results was related primarily to the categorical diagnostic distinction between melancholia and nonendogenous depression. The weakness of the correlation between severity and DST response among melancholic patients was unexpected, but was consistent with the absence of a significant difference in sensitivity between melancholic outpatients and inpatients.

So far, no consistent set of clinical features has been identified which predicts a positive test result among patients with melancholia [12, 13, 34, 35]. *Schlesser* et al. [13, 36] have proposed that genetically distinct subgroups of unipolar depressives can be identified by the DST, but other investigators [35, 37] have not confirmed their report. The results accumulated by many investigators to date indicate that the group of patients diagnosed as melancholic is heterogeneous with respect to neuroendocrine function as assessed by the DST. Some of this heterogeneity may be reduced by the increased DST sensitivity we have described. At present, however, it must be emphasized that a negative or normal DST result will not rule out the clinical diagnosis of melancholia.

The validity of the DST for the diagnosis of melancholia has been discussed at length elsewhere [8]. The test has face validity (high specificity and high predictive value); predictive validity, especially with respect to early relapse after treatment; and construct validity. The neuroendocrine disturbance revealed by the DST indicates dysfunction in the limbic system and hypothalamus, the presumed site of pathology in melancholia. In a recent study we have shown that abnormal DST responses are not related to unusually rapid clearance of dexamethasone by melancholic patients [38].

The limitation of the DST as a laboratory diagnostic test for melancholia has been that its sensitivity was only moderately high (in the range of 40–45%) in most of the studies reported to date [10–18]. Our present results indicate that with a 1-mg dose of dexamethasone with blood samples drawn at 16.00 h and 23.00 h, and with a plasma cortisol criterion value of 5 μg/dl, a sensitivity greater than 65% can be achieved, together with a specificity and diagnostic confidence close to 95%. The routine use of this simple test by internists, family practitioners and psychiatrists who treat depressed patients may help to reduce the diagnostic confusion we have described. The positive identification of melancholic patients by this test, in conjunction with other promising laboratory diagnostic procedures [39], may help also to promote the selective and discriminating use of antidepressant drugs in the treatment of 'depression'.

Summary

We studied 368 psychiatric patients and 70 normal subjects with an overnight dexamethasone suppression test (DST) in order to standardize the test for the diagnosis of patients with melancholia (endogenous depression). Abnormal escape of plasma cortisol concentrations from suppression within 24 h after administration of dexamethasone occurred frequently in melancholic patients, but rarely in other psychiatric patients or in normal subjects. The optimal plasma cortisol criterion concentration above which a DST result may be considered abnormal was 5 μg/dl. The optimal dose of dexamethasone was 1 mg, which gave an overall test sensitivity double that obtained with a dexamethasone dose of 2 mg, without loss of specificity. The most efficient schedule of blood sampling was the combination of samples obtained at 16.00 h and 23.00 h postdexamethasone. These two blood samples detected 98% of the positive tests. The DST procedure used as described above identified melancholic patients with a sensitivity of 67% and a specificity of 96%. The diagnostic confidence for melancholia of a positive DST result was 94%. Based on the plasma cortisol concentrations obtained only at 16.00 h postdexamethasone there was no significant difference in the sensitivity observed among melancholic outpatients and inpatients. Abnormal DST results were not related significantly to age, sex, recent use of psychotropic drugs, or severity of depressive symptoms. There is extensive evidence which validates this practical test for the diagnosis of melancholia.

Acknowledgements

This work was supported by USPHS Grant MH 28294, by research funds from the State of Michigan Department of Mental Health, and by the Mental Health Research Institute, University of Michigan. We thank all the members of our clinical, research and laboratory staffs, particularly *Nancy Barber,* RN, *Catherine Doherty,* RN, *Ruth Anderson-Zabre,* MSW, *Sarah Rawson,* MS, *John Jarrett,* MS and *James Ritchie,* MS. *Peter E. Smouse,* PhD assisted with the development of our data management system.

References

1 Hollister, L.E.: Tricyclic antidepressants. New Engl. J. Med. *299:* 1106–1109 (1978).

2 Hordern, A.; Wheatley, D.: The black cloud. The recognition and treatment of endogenous depression in general practice. Med. J. Aust. *i:* 637–643 (1972).

3 Feighner, J.P.; Robins, E.; Guze, S.B.; Woodruff, R.A.; Winokur, G.; Munoz, R.: Diagnostic criteria for use in psychiatric research. Archs gen. Psychiat. *26:* 57–63 (1972).

4 Spitzer, R.L.; Endicott, J.; Robins, E.: Research Diagnostic Criteria (RDC) for a selected group of functional disorders; 3rd ed. (State Psychiatric Institute, New York 1977).

5 Akiskal, H.S.; Webb, W.L.: Psychiatric diagnosis: exploration of biological predictors, pp. 9–152 (Spectrum, New York 1978).

6 Woodruff, R.A.; Goodwin, D.W.; Guze, S.B.: Psychiatric diagnosis (Oxford University Press, New York 1974).

7 Kendell, R.E.: The classification of depression. A review of contemporary confusion. Br. J. Psychiat. *129:* 15–28 (1976).

8 Carroll, B.J.; Feinberg, M.; Greden, J.F.; Haskett, R.F.; James, N.Mc.I.; Steiner, M.; Tarika, J.: Diagnosis of endogenous depression. Comparison of clinical, research and neuroendocrine criteria. J. affect. Dis. *2:* 177–194 (1980).

9 Carroll, B.J.; Curtis, G.C.; Mendels, J.: Neuroendocrine regulation in depression. I. Limbic system-adrenocortical dysfunction. Archs gen. Psychiat. *33:* 1039–1044 (1976).

10 Carroll, B.J.; Curtis, G.C.; Mendels, J.: Neuroendocrine regulation in depression. II. Discrimination of depressed from non-depressed patients. Archs gen. Psychiat. *33:* 1051–1057 (1976).

11 Carroll, B.J.: Neuroendocrine procedures for the diagnosis of depression; in Garattini, Depressive disorders, Symposium Medicum Hoechst 13, Stuttgart, 1978, pp. 231–236 (Schattauer, New York 1978).

12 Brown, W.A.; Johnston, R.; Mayfield, D.: The 24-hour dexamethasone suppression test in a clinical setting. Relationship to diagnosis, symptoms and response to treatment. Am. J. Psychiat. *136:* 543–547 (1979).

13 Schlesser, M.A.; Winokur, G.; Sherman, B.M.: Hypothalamic-pituitary-adrenal axis activity in depressive illness. Archs gen. Psychiat. *37:* 737–743 (1980).

14 Papakostas, Y.; Lee, J.; Johnson, L.; Fink, M.: Neuroendocrine effects of ECT (Abstract). Soc. Biological Psychiatry, Annual Meeting, 1980.

15 Brown, W.A.; Shuey, I.: Response to dexamethasone and subtype of depression. Archs gen. Psychiat. *37:* 747–751 (1980).

16 Dotti, A.; Bersani, G.; Nocera, F.: Dexamethasone suppression test as factor of differential diagnosis in endogenous depression. Int. Soc. Psychoneuroendocrinology XIth Int. Congr., Florence 1980, abstract 60.

17 Holsboer, F.; Klein, H.; Bender, W.; Benkert, O.: Hypothalamic-pituitary-adrenal activity in a group of 100 heterogenic depressed patients: diagnostic validity and biochemical aspects of the cortisol response to dexamethasone suppression Prog. Neuro-Psychopharmacol., suppl., Abstr. 12th CINP Congr., 1980, abstract 297, p. 180.

18 Nuller, J.L.; Ostroumova, M.N.: Resistance to inhibiting effect of dexamethasone in patients with endogenous depression. Acta psychiat. scand. *61:* 169–177 (1980).

19 Carroll, B.J.; Greden, J.F.; Feinberg, M.: Neuroendocrine disturbances and the diagnosis and aetiology of endogenous depression. Lancet *i:* 321–322 (1980).

20 Spitzer, R.L.; Endicott, J.: Schedule for affective disorders and schizophrenia (New York State Psychiatric Institute, New York 1975).

21 Diagnostic and Statistical Manual of Mental Disorders; 3rd ed., pp. 205–224 (American Psychiatric Association, Washington 1980).

22 Hamilton, M.: A rating scale for depression. J. Neurol. Neurosurg. Psychiat. *23:* 56–62 (1960).

23 Carroll, B.J.; Feinberg, M.; Smouse, P.E.; Rawson, S.G.; Greden, J.F.: The Carroll rating scale for depression. I. Development, reliability and validation. Br. J. Psychiat. *138:* 194–200 (1981).

24 Carroll, B.J.: Clinical application of neuroendocrine research in depression; in Van Praag, Handbook of biological psychiatry. III. Brain mechanisms and abnormal behavior pp. 180–193 (Dekker, New York 1980).

25 Nugent, C.A.; Nichols, T.; Tyler, F.H.: Diagnosis of Cushing's syndrome. Single dose dexamethasone suppression tests. Archs intern. Med. *116:* 172–176 (1965).

26 McHardy-Young, S.; Harris, P.W.R.; Lessoff, M.H.; Lyne, C.: Single-dose dexamethasone suppression test for Cushing's syndrome. Br. med. J. *1:* 740–744 (1967).

27 Murphy, B.E.P.: Some studies of the protein-binding of steroids and their application to the routine micro and ultramicro measurement of various steroids in body fluids by competitive protein-binding radioassay. J. clin. Endocr. Metab. *27:* 973–990 (1967).

28 Vecchio, T.J.: Predictive value of a single diagnostic test in unselected populations. New Engl. J. Med. *274:* 1171–1173 (1966).

29 Galen, R.S.; Gambino, S.R.: Beyond normality. The predictive value and efficiency of medical diagnoses (Wiley, New York 1975).

30 Gaddum, J.H.: Log-normal distributions. Nature, Lond. *156:* 463–466 (1945).

31 Heath, D.F.: Normal or log-normal. Appropriate distributions. Nature, Lond. *213:* 1159–1160 (1967).

32 Carroll, B.J.; Martin, F.I.R.; Davies, B.: Resistance to suppression by dexamethasone of plasma 11-OHCS levels in severe depressive illnesses. Br. med. J. *iii:* 285–287 (1968).

33 Langer, G.; Schonbeck, G.; Koinig, G.; Lesch, O.; Schussler, M.: Hyperactivity of hypothalamic-pituitary-adrenal axis in endogenous depression. Lancet *ii:* 524 (1979).

34 Carroll, B.J.: The hypothalamic-pituitary-adrenal axis in depression; in Davies, Carroll, Mowbray, Depressive illness. Some research studies, pp. 23–201 (Thomas, Springfield 1972).

35 Carroll, B.J.; Greden, J.F.; Feinberg, M.; James, N.Mc.I.; Haskett, R.F.; Steiner, M.; Tarika, J.: Neuroendocrine dysfunction in genetic subtypes of primary unipolar depression. Psychiat. Res. *2:* 251–258 (1980).

36 Schlesser, M.A.; Winokur, G.; Sherman, B.M.: Genetic subtypes of unipolar primary

depressive illness distinguished by hypothalamic-pituitary-adrenal axis activity. Lancet *i:* 739–741 (1979).

37 Clayton, P.: Personal communi. (1980).

38 Carroll, B.J.; Schroeder, K.; Greden, J.F.; Mukhopadhyay, S.; Feinberg, M.; Ritchie, J.; Tarika, J.: Plasma dexamethasone concentrations and cortisol suppression response in patients with endogenous depression. J. clin. Endocr. Metab. *51:* 433–437 (1980).

39 Carroll, B.J.: Implications of biological research for the diagnosis of depression; in Mendlewicz, New advances in the diagnosis and treatment of depressive illness pp. 85–107 (Elsevier, Amsterdam 1980).

Prof. B.J. Carroll, Mental Health Research Institute, University of Michigan, Ann Arbor, MI 48109 (USA)

Adv. biol. Psychiat., vol. 7, pp. 14–25 (Karger, Basel 1981)

Classification: The Bridge over the River Styx

George Winokur

Department of Psychiatry, University of Iowa College of Medicine,
Iowa City, Iowa, USA

The river Styx separated the realm of the living from the resting place of the souls of the dead. The newly dead were ferried over by Charon, the boatman. The final resting place of these souls is an area of the unknown as opposed to the realm of the living. In a sense, this resting place is like the biological substrate of affective disorders, a place about which little is known but much is speculated. In contrast, the realm of the living is similar to the simple process of diagnosis. Much is known but meaning is clouded and outcome is questionable. To traverse the path from diagnosis to pathophysiology, a bridge is needed and that bridge must be composed of classification. This is not to imply that the fortuitous finding of some biological factor might not work backwards and help make sense of classification, but with the dearth of unequivocal advances in biological aspects of affective disorder this seems unlikely.

Diagnosis

The use of systematic criteria, such as the Feighner or the Research diagnostic criteria (RDC) has made it simple to arrive at a reliable method for making diagnosis. Looking at a general population, the Feighner criteria would indicate that about 8% of patients could be diagnosed as having what constitutes a primary major affective illness [1]. Using the RDC, a far higher yield has been found [2]. Table I shows the differences of the use of the two sets of criteria in different places. The Feighner criteria were used in the Iowa 500 study, whereas the RDC were used in the New Haven study. There are differences in the criteria, mostly due to time constraints. To achieve a positive diagnosis with the Feighner criteria requires a longer duration of illness.

Serious researchers can achieve a high degree of reliability. Using the Feighner criteria, *Helzer* et al. [3] reported an average overall agreement for pairs of

Table I. Major depressive illness (including bipolar) in two General population studies

	Number of patients studied	Number of patients with		Morbidity risk	Lifetime ratio
		AD	BZ		
The Iowa 500 study Feighner criteria (illness lasts 1 month)	541	26	344	7.6%	–
The New Haven study Research diagnostic criteria (illness lasts 1–2 weeks)	511	94	–	–	18.4%

AD = Affective disorder; BZ = Number of patients at risk.

Table II. Changes in diagnoses of schizophrenia, mania and depression after follow-up of 1 or more years

	Research diagnosis		
	schizophrenia	manic-depressive illness	depressive illness
Number of patients followed-up 1 or more years	151	66	171
Mean follow-up, years	3	3	5
Follow-up clinical diagnosis			
No further diagnosis	65	33	96
Depressive illness	0	6	56
Manic-depressive illness	0	19	10
Schizophrenia	84	6	5
Other	2	2	4
Unchanged from schizophrenia or affective Disorder in those having a subsequent diagnosis, %	98	76	88
Unchanged from schizophrenia or affective disorder, %	99	88	94

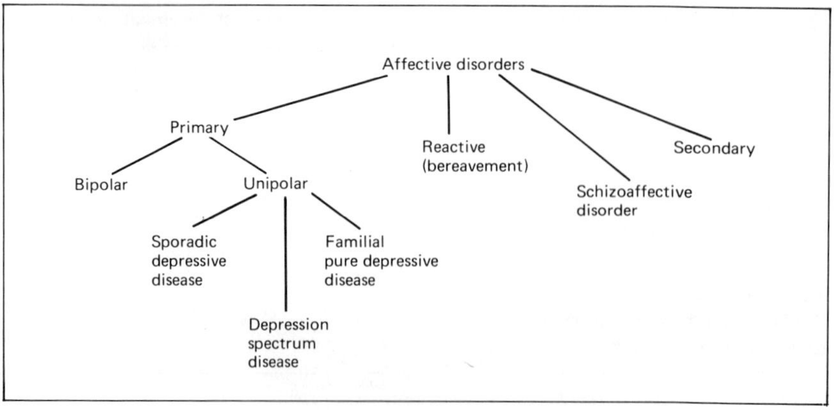

Fig. 1. Breakdown of the affective disorders.

interviewers between 91% and 93%. Kappa values (concordance after correcting for agreement expected due to chance alone) were high, 0.66 to 0.75. For depression, the overall agreement was 86%, the Kappa being 0.55. There seems to be no question but that these criteria are, in fact, stable over time and valid. In 171 depressive patients who were diagnosed using research criteria and followed for a mean of 5 years, 88–94% showed no change in diagnosis [4]. These data are shown in table II.

If fully a fifth of the people in New Haven have a diagnosis of affective disorder of a major sort, one must question whether the criteria are tapping a homogeneous group. Therefore, the most important problem at this point is not the making of a diagnosis but rather the classification of people who fit the diagnosis.

Classification

The usual method for classification in psychiatry depends on the clinical picture and the course of the illness. *Kraepelin* separated dementia precox from manic-depressive illness, but can one consider manic-depressive illness to be a homogeneous entity? On the assumption that it is not, another methodology is needed. Figure 1 gives a breakdown of the affective disorders. All of the entities under the rubric of affective disorders are characterized by showing mania and/or depression and all are syndromes which are characterized by remission. What then is the additional methodology? Clearly it has to do with specific types of family history. First, the affective disorders are divided into primary and second-

Table III. Alcoholism and depression in parents and sibs with and without secondary affective disorder

	n	Median Age	Parents and sibs with alcoholism or depression, %
Hysterics with secondary AD	29	34	8
Hysterics without secondary AD	27	33	9
Alcoholics with secondary AD	38	42	18
Alcoholics without secondary AD	63	48	15

AD = Affective disorder.

ary. This is a clinical separation, secondary affective disorder referring to patients who develop a depressive (but not a manic) episode in the course of another disorder such as alcoholism, anxiety neurosis, schizophrenia, or any psychiatric or serious medical illness. Of interest is that the family history of alcoholism and affective disorder in secondary depressives is similar to the family history of those patients with the same primary diagnosis who have no accompanying secondary depression [5] (table III).

Primary affective disorder is an affective illness (a mania or a depression) which occurs in an individual who has never had another kind of psychiatric illness. There are two other kinds of affective disorders which should stand alone. One of these is bereavement depression, which is an excellent model for reactive depression. Such bereaved and depressed individuals do not have a different kind of family history from those bereaved people who do not develop depressive syndrome [6]. The other is schizoaffective disorder which is diagnosed by the presence in the same individual of both affective symptoms (either manic or depressive) and schizophrenic symptoms, such as delusions or hallucinations of nonaffective quality. There is a vigorous debate over whether schizoaffective disorder is more closely allied with the affective disorders or with schizophrenia. Some have postulated a third and independent psychosis which breeds true in families. Table IV presents data which help elucidate the relationship [7]. There are some nonaffective symptom differences between schizoaffectives (acute schizophrenia) and chronic schizophrenia. Schizoaffectives are more likely to have remitting illnesses and a family history of affective disorder; thus, placing them more likely in the affective disorder category. A recent paper by *Tsuang* [8] presented data on the sibs of schizoaffective probands. There were more ill sibpairs of the schizoaffective-affective disorder type (n = 8) than of the schizoaffective-schizophrenia type (n = 5). Significantly fewer sibpairs of schizoaffective-schizoaffective occurred than would be expected by chance. Therefore, evidence of an autonomous schizoaffective disorder is questionable [8].

Table IV. Differences between acute (schizoaffective) and chronic schizophrenics

	Acute schizophrenics, % (n = 28)	Chronic schizophrenics, % (n = 25)	p
Visual hallucinations	50	12	0.01
Tactile Hallucinations	14	36	n.s.
Any delusion	100	76	n.s.
Passivity	29	16	n.s.
Perceptual disorder	36	8	0.05
Blunted affect	11	40	0.05
Catatonic traits	32	12	0.05
Family history of:			
Schizophrenia, %	18	44	
Affective disorder, %	61	16	
Follow-up			
n	33	119	
Length of follow-up, years	1–2	1–3	
Well at follow-up, %	61	8	

Table V. Family history of affective disorder and EEG in 40 bipolar patients

	Abnormal eeg	Normal egg
Affective disorder in first-, second- and third-degree	2	21
Negative family history	7	6

Differences between two familial subgroups; $p < 0.01$. EEG abnormalities in 10 bipolars, 2 slow waves in left temporal area, 1 slow wave in right temporal area, 7 generalized slow waves. 1 bipolar had an abnormal EEG and a non-affective disorder family history.

Table VI. Postspartum manics versus nonpostpartum female manics

	Postpartum manics	Matched control female manics	Unselected female manics
n	21	21	136
First-rank symptoms, %	62	–	28*
Recurrence of non-postpartum mania in 3-year follow-up	0	6 episodes**	–
Proportion with affective illness in first-degree relatives, %***	19	38	36

* $p < 0.005$; ** $p < 0.05$; *** Differences not significant.

The primary affective disorders may be divided into bipolar and unipolar. Of course, the bipolar patient shows a mania unlike the unipolar patient. The bipolar patient has significantly more mania in his family background. Only 4–5% of unipolar patients become bipolar in follow-up [9, 10]; also many, if not most, of the relatives of bipolar patients who have an affective illness themselves, have only unipolar illnesses. Further, the bipolar patient has more episodes of illness and an earlier onset than the unipolar patient. It is quite possible that the unipolar depressive related to a bipolar patient should be considered as a bipolar patient himself, even though he never shows the mania. As we go from primary to bipolar and unipolar in figure 1, we are now beginning to consider a separation based on family history rather than a separation based on clinical picture and course.

Bipolar Illness

The bipolar family history is very specific in that it contains many family members who have had a mania. One of the things that must be determined is whether or not the bipolar syndrome is homogeneous. Three studies now have shown that patients with EEG abnormalities are less likely to have a family history of bipolar disorder [11]. This does not prove that manics with abnormal EEGs have a different illness than manics with normal EEGs but it is suggestive. Table V shows some familial differences in a recent study. It should not surprise us that there may be heterogeneity within the bipolar group.

Etiological heterogeneity within the bipolar syndrome is suggested also by a study on postpartum mania [12]. 21 bipolar postpartum women were evaluated and compared to both a control group and an unselected group of female manics (table VI). Postpartum bipolar patients had a lower family history of affective illness. It is the same type of finding as we noted in table V as regards abnormal EEGs. Postpartum patients have more Schneiderian first-rank symptoms of schizophrenia. For practical purposes, the most important finding is that the postpartum women had no subsequent nonpostpartum episodes in a 3-year follow-up. If postpartum mania is a different illness from nonpostpartum mania, there may be a reason to treat such a patient differently.

Unipolar Depression

From the New Haven study, one can calculate a ratio of about 14:1 unipolar to bipolar in the general population. In the people who meet the unipolar criteria several illnesses may be represented. Following a familial methodology, it is possible to separate the unipolar disorders into three subtypes [13].

Table VII. Mean ages of onset of depressions separated by family psychopathology

	n	Mean age of onset
Depression spectrum disease	104	34.6
Familial pure depressive disease	86	33.4
Sporadic depressive disease	98	41.3

Kruskal-Wallis one-way analysis of variance comparing groups on age of onset, $\chi^2 = 16.67$, d.f. 2, $p < 0.0002$.

Table VIII. Clinical and course differences between depression spectrum disease and familial pure depressive disease (all female patients)

	Depression Spectrum Disease (n = 105)	Familial pure Depressive Disease (n = 86)	p
History of sexual problems	32%	15%	0.01
History of divorce or separation	20%	6%	0.01
Life-long irritability	20%	8%	0.025
Loss of interest in usual activities	61%	80%	0.005
At least one relapse in follow-up	23%	48%	0.005
Subsequent hospitalization in follow-up	26%	44%	0.025

The first subtype is depression spectrum disease. It is a depression which meets the following criteria. The illness is defined by the presence of at least one alcoholic first-degree relative. Depression (but no mania) may or may not exist in another member of the family. It is notable that this diagnosis is made on the basis of the family history rather than on the basis of the clinical picture. The second subtype is sporadic depressive disease. This kind of patient enters a study with a definite diagnosis of depression but has no alcoholism, antisocial personality, depression, or mania in his family. The third subgroup is familial pure depressive disease. Such a patient has another depressive in the first-degree family relationship but no alcoholism, antisocial personality or mania. By dividing the group of unipolar depressives into these three familial subgroups, it is possible to classify most patients who enter the hospital with unipolar depressions; but some problems exist in this classification. It is possible to have a situation as follows: a proband who has an alcoholic father may suffer from depression. This proband would be considered to have depression spectrum disease. If he has a daughter who has a depression and the daughter becomes a proband, the daughter would have familial pure depressive disease. We do not know the extent of the importance of the distant family history. At this time, we have simply dealt with the psychopathology in primary relatives. This decision may introduce some impurities in the groups.

The three groups are characterized by significant differences. For instance, the ages of onset are quite different; the depression spectrum patients and the familial pure depressive patients being considerably younger at the time of onset of depression than the sporadic depression patients. Table VII presents this material.

There are some differences in premorbid personality and clinical course between groups, particularly depression spectrum disease and familial pure depressive disease (table VIII). The stormy life-style of the depression spectrum patients as well as their relatively infrequent readmissions make them appear similar to patients who are considered to have neurotic depressions. *Kay* et al. [14] in their follow-up of endogenous and neurotic depressives found the same characteristics in the latter. In fact, depression spectrum disease may largely overlap neurotic depression.

One genetic linkage study has been reported on depression spectrum disease [15]. In this study all family members who have alcoholism or antisocial personality or depression are considered as having the illness. All patients with other diagnoses and normals are considered to be well. In a sibpair comparison reflecting 14 families, suggestive data were obtained indicating the possibility of linkage with the α-haptoglobin (α-Hp) gene locus and the 3rd component of complement (C3) locus. Linkage with α-Hp and C3 are both significant. If these results are correct, it would indicate a possibility that there was a linkage between a putative depression spectrum locus and both α-Hp and C3. As both α-haptoglobin and C3 are on different chromosomes, this is impossible. Alternatively, another possibility is that two loci are involved in depression spectrum disease and that we happened to find linkage between the two markers and the two possible depression spectrum loci. This seems unlikely. A third possibility is that one is incorrect, α-Hp or C3. Finally, a fourth possibility is that both are incorrect and are simply chance findings. What is necessary is a replication of these results.

Relevance of Laboratory Findings to Classification

In the best of all worlds, one would like to have laboratory confirmation of the validity of a classification scheme. Certainly, the clinical picture, the course,

Table IX. Abnormal supressor status (nonsupressor) in primary unipolar depressives versus secondary depression [from ref. 16]

Diagnosis	n	nonsupressors, %[1]
Primary unipolar depressives	146	45%
Secondary Depressives	42	0%

[1] Nonsupression is defined as a fasting 8 a.m. serum cortisol >5.0 μg/dl after administration of 1 mg of dexamethasone at 11 p.m. the night before.

Table X. TSH response to TRH in primary versus secondary depressives

Diagnosis	n	Mean Hamilton score	ΔTSH during illness, μU/ml
Primary	13	28.3	5.2
Secondary	5	31.6	10.6[1]

TSH = Thyrotropin; TRH = thyrotropin releasing hormone. After recovery in 11 primary depressives the ΔTSH was 7.3. After recovery in 5 secondary depressives the ΔTSH was 14.4.

[1] Both during illness and after recovery the secondary depressives had significantly higher TSH responses ($p < 0.05$, two-tailed).

the genetic background and treatment results are useful; but, reliable laboratory findings that are specific to specific groups in a classification should support the validity better than anything else. Such confirmatory evidence from biological tests is beginning to emerge.

Separation of primary from secondary depression is quite controversial but it is interesting to note that considerable laboratory evidence supports this difference. A recent study of the dexamethasone suppression test shows that primary unipolar depressives and secondary depressives are markedly different in response [16]. The dexamethasone tests has been extensively used by *Carroll* et al. [17] who have reported that a significant proportion of endogenous depressives have abnormal suppression of serum cortisol when such patients are challenged with a dose of dexamethasone the night before. Table IX shows the differences in proportion of nonsuppressors in the secondary versus the primary depressives. A recent report from Sachar's laboratory compared 13 primary and 5 secondary depressives on the thyrotropin response to thyrotropin releasing hormone infusion (table X). Primary and secondary depressives symptomatologically appeared similar. However, the primary depressives had a much lower re-

Table XI. Differences between primary and secondary depressives on sleep variables [from ref. 19]

	Primary depressives (n = 15)	Secondary depressives (n = 25)	p
REM, latency, min	48.9	78.4	<0.0001
Ratio: REM activity/ sleep time	0.42	0.26	<0.002
Clinical ratings of depression1	24.8	24.0	n.s.

[1] KDS self-rating psychopathological scales.
REM = Rapid eye movement.

Table XII. Abnormal dexamethasone nonsupression in familially defind subtypes of unipolar depression

Diagnosis	Number of subjects	Number of nonsuppressors
Primary depression:		
Familial	28	23 (82%)
Sporadic	35	13 (37%)
Spectrum	23	1 (4%)
Total	86	37 (43%)
Controls	80	0 (0%)

sponse to thyrotropin relearing hormone than the secondary depressives. What is particularly interesting is that after recovery, the differences continued significantly different [18]. Not only are there endocrinological differences between primary and secondary depression, but *Coble* et al. [19] have reported highly significant differences in electroencephalographic sleep variables. Primary depressives had significantly shorter rapid eye movement latencies (time asleep until the onset of the first rapid eye movement period) and higher measures of basic rapid eye movement than secondary depressives (table XI). Delineation of primary versus secondary depression on the basis of 2 nights of EEG sleep was better than 80%. As in other studies, the clinical symptomatology of the primary and secondary depressives was similar. It is quite interesting that such a controversial issue, the primary-secondary dichotomy, is supported by so much laboratory data.

The most striking data in support of a tricotomy of unipolar depressions (depression spectrum, sporadic depressive and familial pure depressive diseases) is the material from the dexamethasone suppression test. We reported a study in which we gave 1 mg of dexamethasone at 11 p.m. and collected blood for a cortisol determination at 8 a.m. [20]. The findings on the unipolar subtypes are presented in table XII. The familial depressives are abnormal suppressors in the majority of cases. A trivial number of depression spectrum patients are abnormal suppressors. The sporadic depressives are midway between the other two groups. Sporadic depressive disease is the one familial diagnosis made because of a lack lf a family history and we know from previous studies that there are many false negatives when one uses a family history method. Interestingly, the three groups look alike as regards symptoms; but we have two illnesses, familial pure depressive disease and depression spectrum disease, which have a different family history, a different pre-existing personality pattern, a different course, same age of onset, and a different response to a laboratory test. It seems entirely reasonable to separate depression spectrum disease from familial pure depressive disease and to consider them autonomous entities.

Discussion

On the basis of clinical symptomatology, course, family background, and emerging biological tests, we are beginning to develop an appropriate classification scheme for the affective disorders. It is noteable that the original separation must be made on the basis of the clinical picture and the course. To this, one may add a sizeable input from the family background. The laboratory data are mainly used for validation of subgroups which have been defined by clinical and genetic homogeneity. Of course, the next step is to subject these groups to studies which are related to both the treatment and etiology.

References

1 Tsuang, M.; Winokur, G.; Crowe, R.: Schizophrenia and affective disorders among first degree relatives of schizophrenics, manics, depressives and surgical controls (submitted for publication).

2 Weissman, M.; Myers, J.: Affective disorders in a US urban community. Archs gen. Psychiat. *35:* 1304–1311 (1978).

3 Helzer, J.; Clayton, O.; Pambakian, R.: Reliability of psychiatric diagnoses. II. The test/retest reliability of diagnostic classification. Archs gen. Psychiat. *34:* 136–141 (1977).

4 Winokur, G.: Diagnostic stability over time in schizophrenia, mania and depression. New Engl. J. Med. *290:* 1027 (1974).

5 Winokur, G.: Family history studies. VIII. Secondary depression is alive and well, and . . . Dis. nerv. Syst. *33:* 94–99 (1972).

6 Clayton, P.; Hallikas, J.; Maurice, W.: The depression of widowhood. Br. J. Psychiat. *120:* 71–78 (1972).

7 Winokur, G.: The use of genetic studies in clarifying clinical issues in schizophrenia; in

Mitsuda, Kukuda, Biological mechanisms of schizphrenia and schizophrenic-like Psychoses (IGAKU Shoin, Tokyo 1974).
8 Tsuang, M.: 'Schizoaffective disorder', dead or alive. Archs gen. Psychiat. *36:* 633–634 (1979).
9 Winokur, G.: Mania and depression: family studies and genetics in relation to treatment; in Lipton, Dimascio, Killam, Psychopharmacology, a generation of porgress (Raven Press, New York 1978).
10 Winokur, G.: Unipolar depression. Is it divisible into autonomous subtypes? Archs gen. Psychiat. *36:* 48–52 (1979).
11 Kadrmas, A.; Winokur, G.: Manic depressive illness and EEG abnormalities. J. clin. Psychiat. *40:* 35–36 (1979).
12 Kadrmas, A.; Winokur, G.; Crowe, R.: Postpartum mania. Br. J. Psychiat. *135:* 551–554 (1979).
13 Winokur, G.; Behar, D.; Valkenburg, C. Van et al.: Is a familian definition of depression both feasible and valid? J. nerv. ment. Dis. *166:* 764–768 (1978).
14 Kay, D.; Garside, R.; Roy, J.; Beamish, P.: Endogenous and neurotic syndromes of depression: A 5 to 7 year follow-up of 104 cases. Br. J. Psychiat. *115:* 389–399 (1969).
15 Tanna, V.; Winokur, G.; Elston, R. et al.: A linkage study of depression spectrum disease: the use of the sib-pair method. Neuropsychobiology *2:* 52–62 (1976).
16 Schlesser, M.; Winokur, G.; Sherman, B.: Hypothalamic-pituitary-adrenal axis activity in depressive illness: its relationship to classification. Archs gen. Psychiat. (in press).
17 Carroll, B.; Curtis, G.; Mendels, J.: Neuroendocrine regulation in depression. II. Discrimination of depressed from non-depressed patients. Archs gen. Psychiat. *33:* 1051–1058 (1976).
18 Asnis, G.; Nathan, R.; Halbreich, U. et al.: TRH tests in depression. Lancet *i:* 424–425 (1980).
19 Coble, P.; Foster, G.; Kupfer, D.: Electroencephalographic sleep diagnosis of primary depression. Archs gen. Psychiat. *33:* 1124–1127 (1976).
20 Schlesser, M.; Winokur, G.; Sherman, B.: Genetic subtypes of unipolar primary depressive illness distinguished by hypothalamic-pictuitary-adrenal axis activity. Lancet *i:* 739–741 (1979).

G. Winokur, MD, Department of Psychiatry, University of Iowa College of Medicine, 500 Newton Road, Iowa City, IA 52242 (USA)

Adv. biol. Psychiat., vol. 7, pp. 26–33 (Karger, Basel 1981)

Neurotransmitter Disturbances and Predisposition to Depressive Illness

Göran Sedvall

Department of Psychiatry, Karolinska Institutet, Stockholm, Sweden

Introduction

Cerebral monoamines were first suspected to play a role in the pathophysiology of depressive syndromes in the early fifties when the clinical and biochemical effects of reserpine were described. This drug was shown to induce depression-like behavioral changes in experimental animals and in patients alike [*Bunney and Davis,* 1965]. Moreover, it also produced a marked reduction in the cerebral concentrations of the monoamines, serotonin, dopamine and noradrenaline [*Pletscher* et al., 1955; *Carlsson and Hillarp,* 1956; *Holzbauer and Vogt,* 1956]. The subsequent findings that monoamine oxidase inhibitors and tricyclic antidepressants alleviate depressive syndromes and that these drugs also induce marked changes in the metabolism of the monoaminergic transmitters in the central nervous system gave further support to the view that cerebral monoamines play a role in the pathophysiology of depressive disorders.

More direct evidence for a role of monoamines in depression was presented by the demonstration of reduced concentrations of the serotonin metabolit, 5-hydroxyindoleacetic acid (5-HIAA), and noradrenaline metabolites, normetanephrine and 3-methoxy 4-hydroxyphenylethylene glycol (MOPEG) in the cerebrospinal fluid (CSF) and urine of depressed patients [*Ashcroft* et al., 1966; *Schildkraut,* 1965]. Evidence for biochemical heterogeneity within groups of depressive patients was later presented by *van Praag* et al. [1973] and *Åsberg* et al. [1976]. These authors demonstrated a bimodal distribution of 5-HIAA in the CSF of depressed patients. However, a number of investigators failed to confirm the presence of low monoamine concentrations in the body fluids of depressive patients. Some of these discrepant results were possibly due to the use of photofluorimetric methods with a relatively low sensivity and specificity for monoamine metabolite determination during the sixties. Another problem in the previous studies was the lack of appropriate control materials.

Mass Fragmentographic Determination of Monoamine Metabolites in the CSF of Healthy Volunteers

The concentrations of the major monoamine metabolites, homovanillic acid (HVA), MOPEG and 5-HIAA, in the CSF are determined by a number of factors. Previous studies indicated that height of subject, age, disturbances in CSF circulation, sex and drug treatment markedly influence the concentrations of these metabolites in the lumbar CSF. Therefore, reference values for the monoamine metabolite concentrations in the CSF from young, healthy subjects had to be established. For this purpose, we recruited 60 volunteers between 18 and 45 years of age. None of the subjects had any previous history of psychiatric disorders or psychoactive drug intake. All were physically and mentally healthy as evaluated by physical examination and rating according to the CPRS scale. CSF was obtained under controlled conditions after one night's bed rest at hospital [*Sedvall* et al., 1980].

Several types of psychiatric disorders are known to be influenced by hereditary factors. Therefore, healthy subjects with a family history of psychiatric disorders in their own or their parents' generations were studied separately from those who did not report such a family history [*Sedvall* et al., 1980]. As shown in figure 1, in subjects lacking a family history of psychiatric disorders, the distribution of 5-HIAA concentration in the CSF was apparently normal. The standard deviation of the distribution in these subjects was significantly lower than in the subjects with a family history of psychiatric disorders. These subjects also appeared to have a multimodal distribution of the 5-HIAA concentration. For HVA and MOPEG we also found significantly greater standard deviations in metabolite concentration in subjects with a family history of psychiatric disorders. Normal reference values for metabolite concentrations were defined on the basis of values within ± 2 SD of the mean value in the group of subjects who did not report a family history of psychiatric disorders. For all the three monoamine metabolites we found significant relationships between deviant concentration of metabolite in the CSF and a family history of psychiatric disorders.

The hospital records of those relatives who had been hospitalized for different types of psychiatric disorders were also examined. It was found that all the subjects with relatives who were treated for depression had 5-HIAA concentrations in the CSF that were below the mean value of the distribution. In fact, subjects with depressive relatives displayed significantly lower 5-HIAA concentrations than subjects who had relatives who were treated for psychotic reactions [*Sedvall* et al., 1980].

In a group of schizophrenic subjects who were shown to have a poor prognosis during a 2-year follow-up, significant relationships could also be demonstrated between high or deviant concentrations of 5-HIAA and HVA in the CSF and a family history of schizophrenia [*Sedvall and Wode-Helgodt,* 1980] (fig. 2).

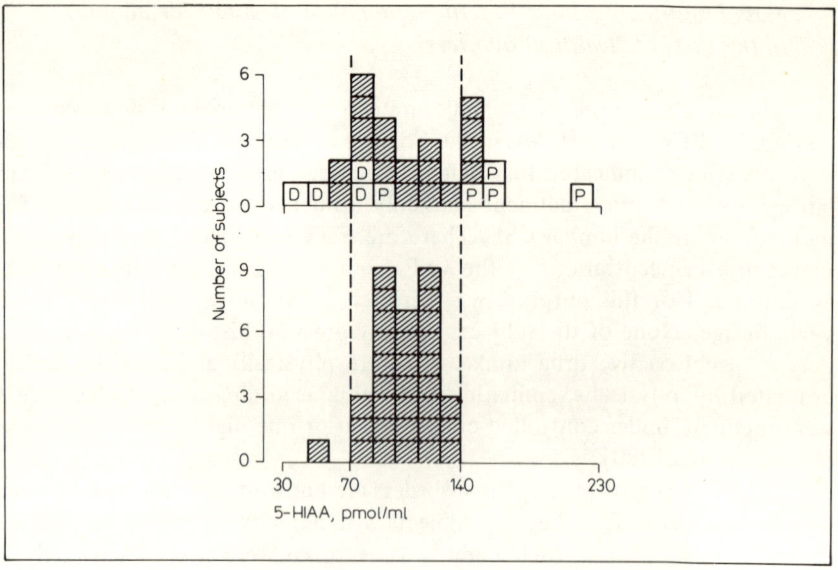

Fig. 1. Distribution of 5-HIAA in CSF of healthy volunteers. *a* Family history of psychiatric disorders. *b* No family history of psychiatric disorders.

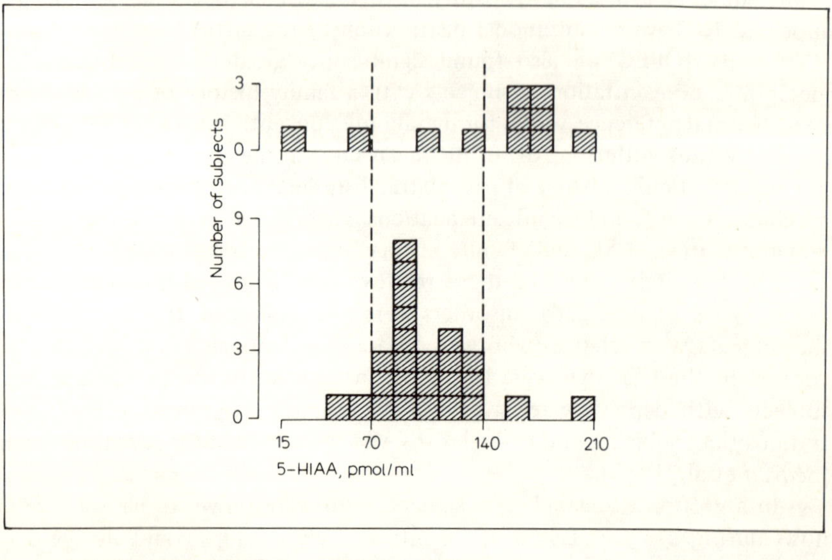

Fig. 2. Distribution of 5-HIAA in CSF of schizophrenic patients. *a* Family history of schizophrenia. *b* No family history of schizophrenia.

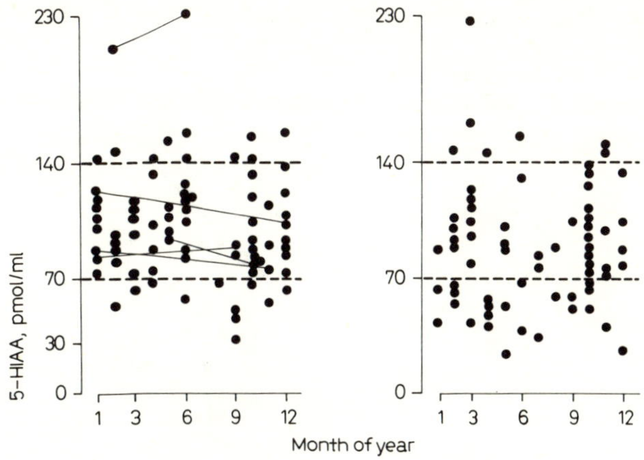

Fig. 3. 5-HIAA concentration in CSF in relation to month of sampling. *a* Volunteers (n = 80). *b* Untreated psychotic patients (n = 67).

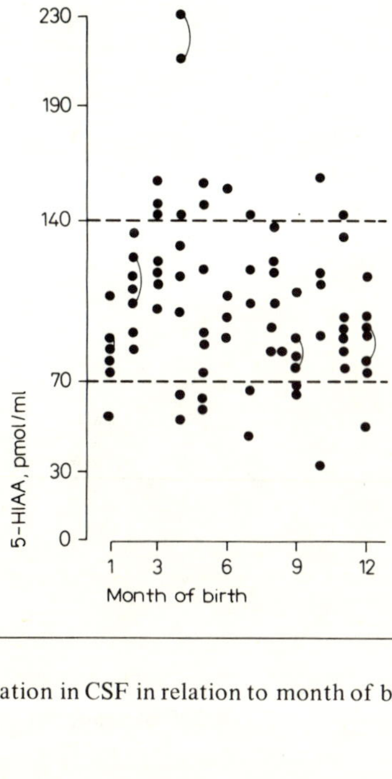

Fig. 4. 5-HIAA concentration in CSF in relation to month of birth in healthy volunteers (n = 80).

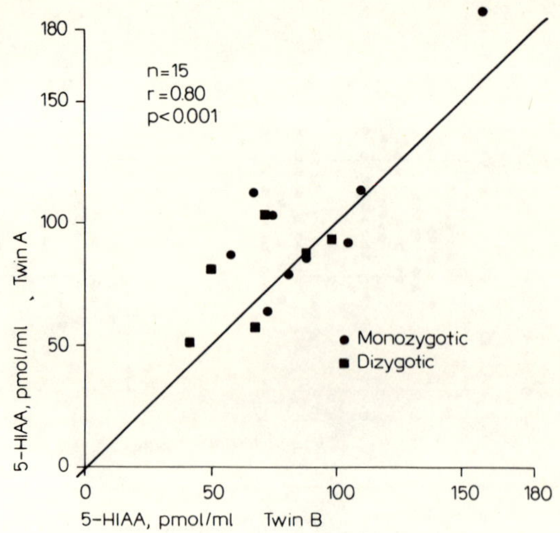

Fig. 5. Correlation between 5-HIAA in CSF in healthy twins.

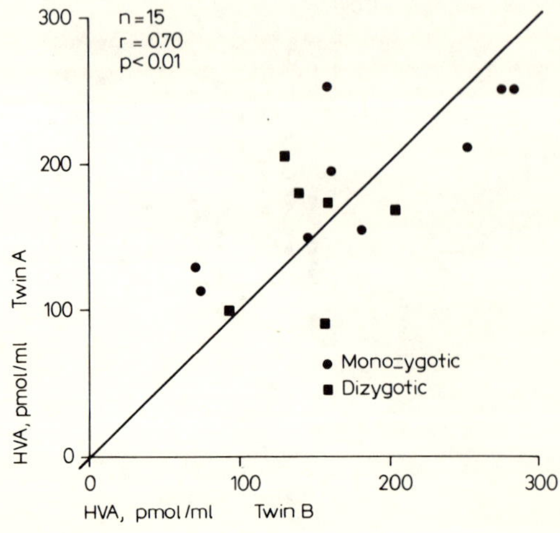

Fig. 6. Correlation between HVA in CSF of healthy twins.

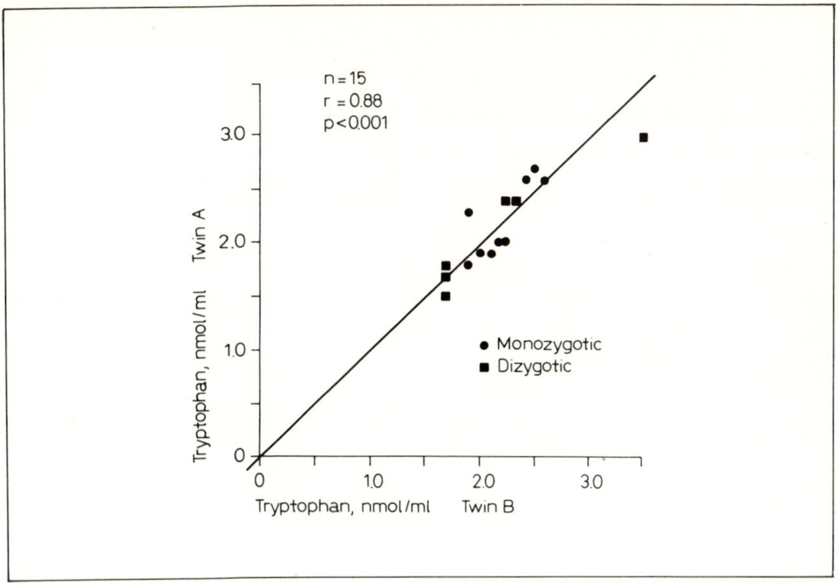

Fig. 7. Correlation between the tryptophan concentration in CSF of healthy twins.

These studies indicated that in both healthy volunteers and schizophrenic subjects there are significant relationships between deviant monoamine metabolite concentrations, predominantly related to serotonin metabolism, and familial occurrence of specific types of psychiatric disorders. Such relationships could not be demonstrated for height or age of the subject, previous hospitalization, previous drug treatment, tryptophan concentration in the CSF, month of CSF sampling (fig. 3) or month of birth of the subject (fig. 4).

The results in the healthy volunteers also indicate that low concentrations of serotonin and dopamine metabolites in the CSF are related to an increased predisposition to depressive disorders among family members.

Monoamine Metabolite Concentrations in the CSF of Twins

The above-described studies indicate a correlation between alterations of cerebral monoamines and familial forms of psychiatric disorders. The occurrence of such relationships could be due to the influence of genetic and/or environmental factors on the development of monoaminergic systems in the central nervous system which may result in an increased predisposition to psychiatric diseases. In order to elucidate the relative roles of genetic and environmental factors in the development of monoaminergic systems in the human brain a study in twins has been initiated [*Oxenstierna and Sedvall,* in preparation]. Mono- and dizygotic male twins, age 20–45, were recruited from the Twin Register of the Karolinska Institute. CSF was obtained from the twins under the same conditions as described above for the healthy volunteers. The preliminary results obtained so far in 15 twin pairs indicate that in both mono- and dizygotic twins there is a high correlation between the metabolite concentrations related to serotonin and dopamine metabolism within the twin pair. This is shown for 5-HIAA and HVA in figure 5 and 6. Such a high correlation was also obtained for tryptophan, the precursor amino acid in the serotonin system (fig. 7). It is of interest that the tryptophan concentration in the plasma was not significantly correlated within the twin pairs.

These results indicate that familial factors determine the concentrations of the transmitter metabolites related to serotonin and dopamine metabolism in the central nervous system. They also indicate a role of familial factors in the steady-state concentration of tryptophan within the central nervous system. So far, the number of twins is too small to allow determination of the relative contributions of genetic and environmental influences to demonstrate similarities in metabolite concentrations.

Summary

The present results indicate that low concentrations of 5-HIAA and HVA in the CSF may be related to an increased risk of developing depressive disorders in family members. The results also indicate the importance of familial factors in the development of monoaminergic mechanisms in the human central nervous system. It seems possible that familial predisposition to depressive disorders may be related to specific familial influences on central serotonin and dopamine metabolism.

Acknowledgement

The present studies were supported by grants from the Swedish Medical Research Council (No. B80 21X-03560), the National Institute of Mental Health, Bethesda, M., USA (MH 27254-03), F. Hoffman-La Roche & Co., Basel, Switzerland, 'Magnus Bergvalls Stiftelse' and Karolinska Institutet.

References

Åsberg, M.; Thorén, P.; Träskman, L.; Bertilsson, L.: Serotonin depression – a biochemical subgroup within the affective disorders? Science *191:* 478–480 (1976).

Ashcroft, G.W.; Eccleston, D.; Crawford, T.B.B.; Sharman, D.F.; MacDougall, E.J.; Stanton, J.B.; Binns, J.K.: 5-Hydroxyindole compounds in the cerebrospinal fluid of patients with psychiatric or neurological diseases. Lancet *i:* 1049–1052 (1966).

Bunney, W.E.; Davis, J.M.: Norepinephrine in depressive reactions. Archs gen. Psychiat. *13:* 483–494 (1965).

Carlsson, A.; Hillarp, N.-Å.: Release of adrenaline from the adrenal medulla of rabbits produced by reserpine. K. fysiogr. Sällsk. Lund Förhandl. *26:* 8–9 (1956).

Holzbauer, M.; Vogt, M.: Depression by reserpine of the noradrenaline concentration in the hypotalamus of the cat. J. Neurochem. *1:* 8–11 (1956).

Pletscher, A.; Shore, P.A.; Brodie, B.B.: Serotonin release as a possible mechanism of reserpine action. Science *122:* 374–375 (1955).

Praag, H.M. van; Korf, J.; Schut, T.: Cerebral monoamine and depression. Archs gen. Psychiat. *28:* 827–831 (1973).

Schildkraut, J.J.: The catecholamine hypothesis of affective disorders: a review of supporting evidence. Am. J. Psychiat. *122:* 509–522 (1965).

Sedvall, G.; Fyrö, B.; Gullberg, B.; Nybäck, H.; Wiesel, F.-A.; Wode-Helgodt, B.: Relationships in healthy volunteers between concentrations of monoamine metabolites in cerebrospinal fluid and family history of psychiatric morbitidy. Br. J. Psychiat. *136:* 366–374 (1980).

Sedvall, G.; Wode-Helgodt, B.: Aberrant monoamine metabolite levels in CSF and family history of schizophrenia. Archs gen. Psychiat. *37:* 1113–1116 (1980).

G. Sedvall, MD, Department of Psychiatry, Karolinska Hospital, S–104 01 Stockholm (Sweden)

Adv. biol. Psychiat., vol. 7, pp. 34–41 (Karger, Basel 1981)

Amines, Aging and Affective Disorders[1]

C.G. Gottfries

University of Göteborg, Department of Psychiatry and Neurochemistry
St. Jorgen's Hospital, Hisings Backa, Sweden

Amines and Affective Disorders

A relation is supposed to exist between behavioral disturbances in affective disorders on the one hand and disorders of the central monoamine metabolism on the other. Relevant data have been obtained from cerebrospinal fluid (CSF) studies, postmortem examinations of the brain and studies of the psychological effects produced by drugs exerting a more or less specific influence on the central monoamine metabolism. I will not go into details about these studies, but I will briefly mention that depression is associated with a decreased turnover of 5-hydroxytryptamine (5-HT) and noradrenaline (NA). The decreased monoamine turnover reflects a decreased transmission in monoaminergic synapsis and this phenomenon is linked to the occurrence of depressive symptoms.

The disorders of the monoamine metabolism are supposed to play a role in the pathogenesis of depressions. The reason for stating this is that when the disorders are controlled by pharmacological treatment there is an improvement of the depressive symptoms.

Affective Disorders in Old People

It is well known that old people need not only physical but also psychiatric care. The frequency of first admissions to psychiatric hospitals increases very much after the age of 70. There are almost 100% more first admissions among old age-groups than among younger people. In figure 1, the first admissions to

[1] This investigation was supported by grants from the Swedish Medical Research Council No. B81-21X-05002-05, 'Fredrik och Ingrid Thuring's Stiftelse, Stiftelsen Hjalmar Svenssons Forskningsfond'.

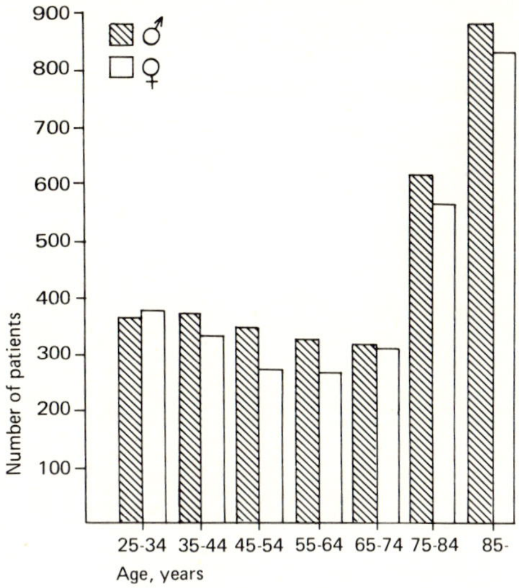

Fig. 1. First admissions in psychiatric hospitals in Sweden 1964 related to age and sex [Medicinalstyrelsen, 1967].

psychiatric hospitals in Sweden in 1964 are presented, related to age and sex [Medicinalstyrelsen, 1967].

It is very often assumed that the increased number of aged people in psychiatric hospitals represents patients with dementia disorders. Investigations in the United Kingdom have shown, however, that 50% of the geriatric patients admitted to psychiatric hospitals suffer from affective disorders [*Kay* et al., 1964]. Only 25% of the aged patients admitted to hospital have organic dementia. In Sweden, it was also found that the number of hospitalized geriatric patients with affective disorders was increasing. Figure 2 shows the number of patients with affective psychoses per 100,00 inhabitants on 1 day in 1964, in Swedish psychiatric hospitals in relation to age and sex. The increased frequency of suicides, especially among old men, also supports the assumption of a higher incidence of affective disorders in old age.

It may be questioned whether affective disorders are correctly diagnosed in old age-groups. From figure 2 it is evident that the frequency of manic depressive disorders and involutional melancholia is much higher in women than that in men. In British investigations it has been shown that affective disorders in old age are about as frequent in women as in men. A study of the number of pre-

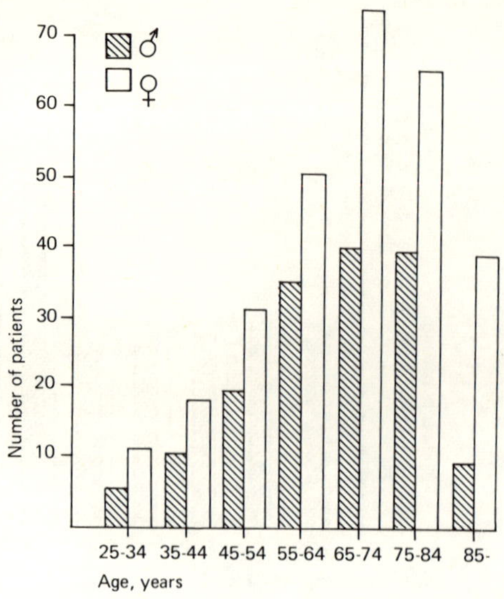

Fig. 2. Number of patients in psychiatric hospitals with affective psychoses on a certain day in 1964 per 100,000 inhabitans [Medicinalstyrelsen, 1967].

scriptions for antidepressives in ambulatory services shows that there is a peak in the age-group 65–74 years and here, women predominate over men (fig. 3) [*Grimsson* et al., 1977]. If it is true that the distribution between women and men is fairly equal, these results indicate that there are many men with undiagnosed depressive disorders.

It is evident that there are diagnostic problems when studying affective disorders in the elderly. The clinical picture when old people are depressed is very often dominated by apathy, inhibition and memory disturbance. These symptoms may be mistaken for symptoms of an organic illness. The affective syndromes which come under discussion in old age are usually: manic depressive illness, senile melancholia, pseudodementia-depression, masked depression, organic depression and agitated depression.

Manic depressive illness is a life-long disease and will therefore also be found in late life. The diagnosis is easy, as the disease often has its onset in early life. A careful case history is sufficient to delimit this group.

Senile melancholia, pseudodementia-depression and masked depression are generally restricted to the old age-group of depressed patients. In senile melancholia, hypochondriacal ideas and delusions may appear; in pseudodementia, the

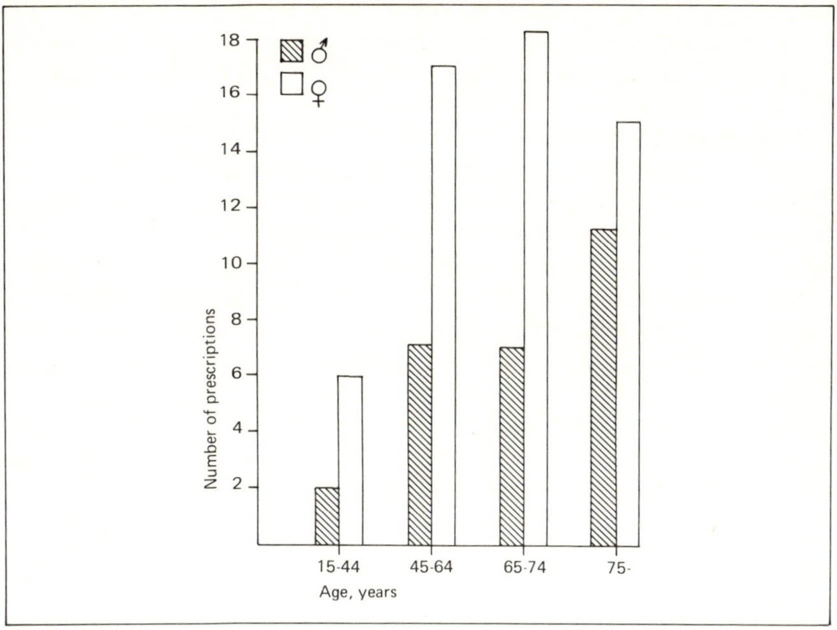

Fig. 3. Number of prescriptions of antidepressive drugs related to age in Sweden in 1975.

picture is dominated by intellectual inhibition and in masked depression or atypical depression there are almost no symptoms of reduced level of mood but the patient shows neurotic and/or physical symptoms. Careful investigations, however, indicate, that these groups of depression should also be included in the group of endogenous depressions [*Angst*, 1966; *Vanggaard*, 1976].

Older patients and patients with organic dementia very often show symptoms of disturbed mood. Whether normal aging involves symptoms of disturbed mood may be discussed but it is evident that dementias cause mood disorders. A study of an unselected group of dementia patients revealed that more than 50% of the patients showed symptoms of depression [*Gottfries*, 1971].

Agitated depressions are those depressions where agitation dominates the clinical picture. This is believed to represent a form of reactive depression where psychogenic and sociogenic factors are of etiological importance.

Investigations by *Blumenthal* [1971] show that there is a higher degree of heterogeneity in affective disorders in old as compared to young patients. This is true for etiology, symptomatology and also progress under treatment. *Blumenthal* [1971] concludes that when discussing etiology one must think of a chain of factors. Psychogenic and sociogenic factors may precipitate the disease but biochemical disturbances, possibly related to the aging process, may make the individual vulnerable to affective disorders. It is natural to assume that retirement, bereavements and somatic illness are factors which cause depression in the elderly. According to *Post* [1976], the relation between psychogenic factors and depression in old age is not proven. He investigated 92 patients with depression in old age. The patients were rated with the Newcastle Diagnostic Index Score. This rating scale is supposed to differentiate endogenous from exogenous depressions but in the old people there was a wide overlap between these two groups. It was shown that some of the psychogenic factors investigated were equally frequent in the endogenous and exogenous groups. It must, indeed, be further investigated to what extent sociogenic and psychogenic factors, somatic illness and treatment with drugs are of etiological importance in affective disorders in the elderly.

As biochemical changes have been assumed to occur in affective disorders, it is of interest to discuss findings concerning the changes in the metabolism of brain monoamines related to aging and dementia disorders.

Aging and Neurotransmitter Metabolism

In normal aging there is an age-related decrease in the levels of dopamine and homovanillic acid at least in the age distribution above 50 [for review see *Gottfries,* 1980]. The results of biochemical investigations in normal aging are summarized in table I. As is evident, the levels of noradrenaline and 5-hydroxytryptamine also decrease with age. The activity of the enzymes monoamine oxidase B and superoxide dismutase increased with age [*Winblad* et al., 1980].

In CSF studies, the levels of the metabolites, 5-hydroxyindoleacetic acid homovanillic acid were shown to increase with age [*Gottfries,* 1980].

The findings from postmortem brain material and CSF do not seem to be in agreement. We interpret the findings in the following way: the reduced levels of monoamines and monoamine metabolites in the brain tissue indicate a reduced monoamine turnover with age. The higher levels of metabolites in CSF indicate that the excretion of these acid metabolites is reduced with age.

There appear to be only quantitative differences between the biochemical changes in normal aging and in Alzheimer type dementia (table I). It is of interest, however, that the end metabolite of noradrenaline metabolism, 3-methoxy-4-hydroxy-phenylglycol, is increased in Alzheimer type dementia. Noradrenaline is a stress hormone and the increased levels of 3-methoxy-4-hydroxy-phenylglycol may indicate an increased noradrenaline turnover.

If the metabolism of 5-hydroxytryptamine and noradrenaline is of importance in the pathogenesis of affective disorders, then it is very likely that the biochemical changes in normal aging and in Alzheimer type dementia disorders en-

Table I. Changes in biological systems related to age or to Alzheimer type dementia according to postmorten investigations in man

Type of change	Normal aging	Alzheimer type dementia
Choline acetyl-transferase	↓	↓ ↓
Muscarine receptors in brain tissue	↓	–
Tyrosinehydroxylase	↓	
Dopadecarboxylase	↓	↓ ↓
Dopamine	↓	↓ ↓
Homovanillic acid	↓	↓ ↓
Noradrenaline	↓	↓ ↓
3-Methoxy-4-hydroxy-phenylglykol	–	↑
5-Hydroxytryptamine	↓	↓ ↓
5-Hydroxyindoleacetic acid	–	↓
Monoamine oxidase A	–	–
Monoamine oxidase B	↑	↑ ↑
Superoxidase dismutase	↑	↑ ↑

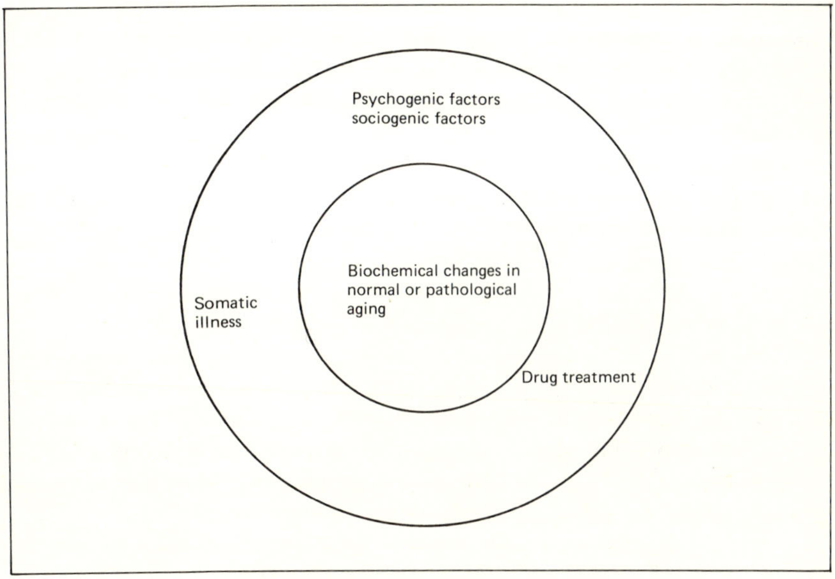

Fig. 4. Etiological factors of importance in affective disorders in old age.

hance the vulnerability to affective disorders in old and in demented patients (fig. 4). We have preliminary data which indicate that in dementia disorders with onset in very old age, i.e. senile dementias, 5-HT metabolism is more severely affected. As *Robinson* et al. [1972] have shown, there is increased monoamine oxidase activity in old age. In our investigation we have been able to show that it is the monoamine oxidase B form which is increased [*Gottfries* et al., 1975]. In disorders of Alzheimer type dementia there is a still greater increase in monoamine oxidase B activity. The hypothesis that this increased monoamine oxidase activity may have a pathogenetic importance in affective disorders has been put forward. We think, however, that one explanation for this increased monoamine oxidase activity may be that the monoamine oxidase B form is restricted to extraneuronal tissues, for instance the glial cells. In old subjects and in dementia disorders, there is glial proliferation, and the increased monoamine oxidase B activity may only reflect an increase in this part of the tissue.

As is evident from table I, there is also reduced activity in the enzyme choline acetyltransferase, both in normal aging and in Alzheimer type dementia. This enzyme synthesizes the important transmitter, acetylcholine. No less than three research groups in the United Kingdom have shown that the activity of choline acetyltransferase is reduced in Alzheimer type dementia. We have been able to confirm these findings [for a review see *Gottfries,* 1980]. There is no hypothesis which includes a disturbed acetylcholine metabolism in the pathogenesis of affective disorders. This finding is nevertheless of interest when discussing depressions. In the management of depression, one important treatment is the use of tricyclic antidepressants. These drugs improve affective disorders in 60–75% of patients. It is, however, well known to clinicians that this kind of treatment may cause problems in elderly patients. Old patients are more sensitive to the side-effects of tricyclic antidepressants and these side-effects sometimes disturb them so much that the treatment cannot be completed. It can be assumed that the sensitivity of old patients to anticholinergic drugs is due to the age-related reduced activity in the acetylcholinergic system.

Treatment Aspects

The biochemical findings can be interpreted as indicating an increased vulnerability to depression in old age due to reduced activity of the serotoninergic system. It is also of importance to keep in mind that the acetylcholinergic system shows reduced activity in the brain of old people. In view of this, drugs with a selective inhibitory effect on 5-hydroxytryptamine reuptake in the synaptic space may be of special interest. Some of these drugs also seem to have few or no anticholinergic side-effects. This type of drugs, at least theoretically, should be the drugs of choice when treating affective disorders in the elderly.

References

Angst, J.: Zur Ätiologie und Nosologie endogener depressiver Psychosen. Eine genetische, soziologische und klinische Studie. Monogr. Neurol. Psychiatr. (Berlin) *112:* 1–118 (1966).

Blumenthal, M.D.: Heterogeneity and research on depressive disorders. Archs gen. Psychiat. *24:* 524–531 (1971).

Gottfries, C.G.: Depressionstillstånd i hög ålder. I. Aktuella problem inom åldringsvården. 5. Symp. arrangerat av Pharmacia, Uppsala 1971 Pharmacia *1971:* 31–38.

Gottfries, C.G.: Amine metabolism in normal ageing and in dementia disorders, in Roberts, Biochemistry of dementia, pp. 213–234 (Wiley, New York 1980).

Gottfries, C.G.; Oreland, L.; Wiberg, Å.; Winblad, B.: Lowered monoamine oxidase activity in brains from acoholic suicides. J. Neurochem. *25:* 667–673 (1975).

Grimsson, A.; Idänpään-Heikkilä, J.; Lunde, M. et al.; Förbrukningen av psykofarmaka i Finland, Island, Norge och Sverige. Nord. Med. *92:* 49–54 (1977).

Kay, D.W.K.; Beamish, P.; Roth, M.: Old age mental disorders in Newcastle upon Tyne. I. A study of prevalence. Br. J. Psychiat. *110:* 146–158 (1964).

Medicinalstyrelsen 1967 missing

Post, F.: Diagnosis of depression in geriatric patients and treatment modalities appropriate for the population; in Gallant, Simpson, Depression: behavioral, biochemical, diagnostic and treatment concepts, pp. 205–231 (Spectrum, New York 1976).

Robinson, D.S.; Davis, J.M.; Nies, A.; Colburn, R.W.; Davis, J.N.; Bourne, H.R.; Bunney, W.E.; Shaw, D.M.; Coppen, A.J.: Ageing, monoamines, and monoamine oxidase levels. Lancet *i:* 290–291 (1972).

Vanggaard, T.: Atypical endogenous depression. Diagnostic criteria. Acta psychiat. Scand., suppl. 267 (1976).

Winblad, B.; Adolfsson, R.; Aquilonius, S.M.; Carlsson, A.; Eckernäs, S.Å.; Gottfries, C.G.; Marklund, S.; Nordberg, A.; Oreland, L.; Svennerholm, L.; Wiberg, Å.: Biogenic amines and related enzymes in normal aging, senile dementia and chronic alcoholism. Proc. CINP-Congress, Göteborg 1980; in Usdin, Riederer, Transmitter biochemistry of human brain tissue (Macmillan, in press).

C.G. Gottfries, MD, University of Göteborg, Department of Psychiatry and Neurochemistry St. Jorgen's Hospital, S–422 03 Hisings Backa (Sweden).

Adv. biol. Psychiat., vol. 7, pp. 42–70 (Karger, Basel 1981)

Central Peptide Function in Affective Illness: Arginine Vasopressin as a Model System

Philip W. Gold[a], *Robert M. Post*[b], *Herbert Weingartner*[b], *Frederick K. Goodwin*[a]

[a] Clinical Psychobiology Branch, and [b] Biological Psychiatry Branch, National Institute of Mental Health, Bethesda, USA

Introduction

It is firmly established that integrative communication within the central nervous system (CNS) proceeds not only by classical transsynaptic neurotransmission, but also by interactions between brain and peptide hormones synthesized entirely in brain by the endocrine hypothalamus. Almost all of the hypothalamic peptide hormones identified so far occur in brain beyond the boundaries of the hypothalamus and are conveyed from their sites of synthesis by either axon transport, local bloodstream circuits, or the cerebrospinal fluid (CSF) communication pathway [1, 2]. Since these peptides function as traditional hormones that convey chemical information over some distance but are produced and distributed primarily in the CNS, it is customary to designate them as neurohormones. Compared with classic neurotransmitters, neurohormones exert effects on brain function at much lower concentrations and for much longer periods of time [2]. These properties, in association with their widespread distribution, render peptide neurohormones well-suited to influence the activity of multiple nerve cell aggregates organized to coordinate complex behavioral and physiological processes.

The study of arginine vasopressin (AVP) function in experimental animals has contributed significantly to current theories of the role of neurohormones in the CNS. For example, AVP was one of the first neurohormones demonstrated to be widely distributed in the CNS [1, 4], to function as a long-acting neuromodulator at minute concentrations [3], and to exert complex behavioral effects [4]. The purpose of this paper is to illustrate the study of central peptide function using AVP as a model system in clinical psychiatric research. To accomplish this task, we first review and integrate data from studies in experimental animals that demonstrate the concept that a single peptide such as AVP can influence virtually all of the CNS processes altered in a well-characterized syndrome such as primary affective disorder. Next, we introduce strategies not previously used in psychiatry to study comprehensively the functions of this peptide in clinical populations. Finally, we present data from a coordinated series of clinical studies that suggest a role for AVP in distrubances of human behavior with possible implications for biochemical diagnosis and treatment.

Central AVP Function and the Syndrome of Primary Affective Disorder

The syndrome of affective illness is characterized not only by alterations in mood, but frequently also is accompanied by changes in memory function [5], the quality and timing of rapid eye movement (REM) sleep [6, 7], processes that relate to pain and reward [8], the periodicity of biological rhythms [9], the status of fluid and electrolyte balance [10], and specific changes in neuroendocrine function [11, 12]. Studies in experimental animals have shown that AVP may influence each of these processes.

The role of AVP in the regulation of memory has been well described by *de Wied* et al. [13]. Rats with a hereditary absence of vasopressin secretion have difficulty in maintaining learned avoidance tasks, a deficiency promptly reversed by the administration of AVP or its analogs [13]. Experimentally induced amnesia of avoidance learning, either after puromycin treatment [13] or intrathecal administration of AVP antiserum [14], is also reversed by AVP administration. Recently, AVP has also been shown to enhance the retention of learned passive avoidance responses in normal rats [15]. It has been suggested that AVPs effect on memory function reflects the peptide's capacity to influence norepinephrine turnover in several hypothalamic nuclei identified as important in the process of memory consolidation [16]. Modification of behavior by regulation of nonadrenergic systems links AVP function and the postulated role of catecholamines in the pathophysiology of affective illness.

The effect of AVP on memory function may also relate to its regulation of hippocampal theta frequencies during REM sleep [17], a stage of sleep that may be important for long-term memory. Drugs that enhance memory produce changes in REM sleep similar to those induced by AVP [18]. Administration of AVP was shown to modulate the onset of REM sleep [22], clearly abnormal in depressed patients [7].

Recent work ties AVP function to that of the endogenous opiate peptides, particularly the processing of information relating to reward and pain phenomena. Thus, AVP significantly alters heroin self-stimulation behavior [19], and the rate of tolerance development during opiate administration [20]. Physiologically, AVP and the opiates appear to participate in a regulatory loop, in which each influences the other's release. For example, both synthetic narcotics [21], and beta-endorphin [22] cause antidiuresis via AVP release, while AVP was shown to release opiates both in vitro [23] and in vivo [24].

A specific neuromodulator has not been implicated in the alterations in biologic rhythms reported in patients with unipolar and bipolar illness. It is noteworthy that AVP is a major constituent of the suprachiasmatic nucleus [25, 26], identified as the locus of the endogenous biological clock [27]. Of additional significance is the fact that the few drugs reported to have profound effects of AVP secretion and function are among the very few pharmacologic agents that influence the periodicity of free-running rhythms. For example, lithium, ethanol and deuterium dioxide all slow the periodicity of free-running rhythms [28] and inhibit overall AVP function [29, 30].

AVP, known primarily as the antidiuretic hormone, influences free water clearance at the renal level [24], and thus maintains the osmolality of body fluids. Alterations in fluid and electrolyte balance have been widely reported in depressed patients [10, 31] and could relate to alterations in central peptidergic modulation of water and salt metabolism [32]. By modulating the permeability of blood-brain barrier capillaries, AVP also seems to exert a direct effect on the fluid and electrolyte status of the CNS [33] and influences the rate and composition of CSF formation by the choroid plexus [34].

AVP was detected in hypophyseal portal blood in concentrations sufficient to promote growth hormone [35] and corticotropin (ACTH) release [36]. Although it is postulated that 'modified' AVP is the primary corticotropin-releasing factor [37], many investigators feel that other hypothalamic peptides play an equally important role in the regulation of ACTH secretion [38]. No studies exist exploring the relationship between alterations in AVP function and reported changes in growth hormone and ACTH regulation in patients with affective illness [11, 12].

It is noteworthy that some of the drugs in the treatment of affective illness exert a profound effect on AVP secretion or action. For instance, lithium and carbamazepine, two drugs shown to possess both antidepressant and antimanic properties [39, 40], effect AVP function and activity in experimental animals and humans [41, 42]. We postulate that their therapeutic efficacy may be partially due to their effect on central AVP function.

Structural and Functional Aspects of AVP Secretion: Rationale of the Strategies to Study AVP Function in Humans

AVP is synthesized entirely within the CNS, specifically by the supraoptic, paraventricular, and suprachiasmatic nuclei. In humans, two major secretory systems convey AVP to sites where it can be directly sampled. The supraoptico-hypophyseal tract transports AVP to the posterior pituitary where it is stored and subsequently released into the systemic circulation to modulate renal free water clearance. A secretory system from the paraventricular nucleus to the third ventricle may mediate cognitive and other central effects of AVP. Long axons also emanate from the suprachiasmatic nuclei to the median eminence (also supplied by the supraoptic nucleus), and to diverse extrahypothalamic regions such as septum, stria terminalis, amygdala, periaqueductal gray, and habenula. AVP synthesized in the suprachiasmatic nucleus and transported to the hypophyseal portal system is thought to influence the synchronization of neuroendocrine circadian rhythms, while AVP synthesized in the suprachiasmatic nucleus and transported through the extrahypothalamic pathways may influence functions such as emotional behavior and analgesia.

Three physiological variables, osmolality, hemodynamic factors (particularly hypotension), and emesis regulate AVP secretion into both the CSF and plasma. Osmolality is by far the most important of these variables, though emesis or large changes in blood pressure or blood volume can influence CSF and plasma AVP levels. The cells that mediate these physiological variables are separate from the neurons that synthesize and release AVP. These cells coney information to the neurosecretory neuron by both neurotransmitters (norepinephrine, dopamine, serotonin) and peptidergic neuromodulators (angiotensin, endorphins). Little is known about which peptides or neurotransmitters are involved in the translation of specific physiologic stimuli into excitation or inhibition of AVP secretion, or the site or synapse where these neuromodulators function.

We decided to measure CSF AVP levels as well as the plasma AVP response to an osmotic stimulus, hypertonic saline infusion. The latter paradigm was chosen since osmolality is the major physiologic regulator of AVP secretion in both CSF and plasma; moreover the plasma response to changes in osmolality is entirely; a centrally mediated phenomenon despite the fact that a measurable output is obtained in the periphery. We also examined the effects of psychoactive agents on the AVP response to hypertonic saline in the evaluation of neuromodulatory factors that influence the hypothalamic osmoreceptor in humans, and to evaluate the possibility that some drugs exert therapeutic efficacy due to their effect on AVP secretion.

In addition to these measurements of AVP in biological fluids during basal conditions and after physiological stimulation, we have also administered a peptide analog of AVP to both depressed patients and controls to examine its effects on cognition, behavior, and electroconvulsive treatment (ECT)-induced retrograde amnesia. A large number of AVP analogs were shown to have cognitive effects in experimental animals; we selected 1-desamino-8-*D*-arginine vasopressin (DDAVP), an analog which possesses a long half-life but is devoid of pressor activity, thus permitting sustained administration of large doses.

Methods

The patients were hospitalized on the clinical research units of the NIMH and met Research Diagnostic Criteria for major affective disorder [43]. Volunteer subjects were studied on the same clinical research units. During each hospital day, patients were rated twice daily under double-blind conditions by a trained nursing research team using the NIMH scale for global items of depression, anxiety, psychosis, mania, and anger [44]. Both patients and volunteers adhered to a controlled monoamine diet for at least 10 days and, unless otherwise noted, all subjects received no active medication for at least 3 weeks prior to the study. The subjects chosen had a normal fluid and electrolyte status, blood pressure within normal limits and were free of clinical signs of volume depletion or volume overload. Specifically, any subject with hypertension (BP greater than 130/190), hypotension (BP less than 90/60), or labile blood pressure (day-to-day variation in the mean of three blood pressure recordings greater than 10%) was excluded from the study. Any subject showing clinical signs of volume depletion or dehydration (hypernatremia, decreased urine sodium, weight loss greater than 2 kg) or volume overload (edema, weight gain of greater than 2 kg from the mean weight) within 2 weeks of any clinical study was excluded. Three experimental paradigms were used: (1) measurement of CSF AVP; (2) plasma response to intravenous hypertonic saline, and (3) assessment of cognitive and behavioral responses to DDAVP treatment. These paradigms will be discussed separately, giving a description of the subjects included in each study, additional criteria for inclusion or exclusion of subjects in a given study, and the exact details of each procedure.

CSF AVP Studies

CSF AVP was obtained in drug-free patients and controls at 9.00 a.m. All patients were kept NPO after midnight and were at bedrest for at least 3 h before the lumbar puncture. Blood pressure was measured every 5 min for 30 min prior to the lumbar puncture. The subject was excluded from the study if there was a fall in mean blood pressure of more than 10 mm from one recording to another, if nausea or other evidence of vasovagal reaction was seen, or if the subject could not be maintained at bedrest for at least 3 h before the procedure. To test for the possible effects of increased activity and sleeplessness on CSF AVP, 2 euthymic drug-free patients with primary affective disorder were asked to arise at 3.00 a.m. and to participate in relatively strenuous physical exercise until 6.00 a.m.; from 6.00 a.m. until 9.00 a.m. they remained at complete bedrest to meet criteria for inclusion in the study.

Samples were immediately placed in dry ice and frozen at -70 °C. For CSF AVP and osmolality determinations, 1.5 cm^3 of an aliquot representing the gradient from 21 to 26 cm^3 was used. *Luerssen and Robertson* [45] reported that the concentration of AVP is relatively uniform throughout, but use of the same aliquot from each subject minimizes any possibility of artifacts introduced by possible gradients for CSF AVP.

74 subjects, 32 normal volunteers and 42 patients, participated in the CSF AVP study. 21 drug-free nonpsychotic depressed patients were studied; of these, 9 were unipolar and 12 were bipolar. 1 unipolar and 1 bipolar patient was studied during psychotic depression. 8 bipolar patients were studied in the manic phase. An additional 2 unipolar and 7 bipolar patients were studied in the euthymic phase.

Osmoregulation of AVP – Hypertonic Saline Infusion

For studies of the plasma AVP response to intravenous hypertonic saline, patients were kept NPO after midnight and until termination of the experimental procedure. They were awakened at 7.00 a.m. and allowed to move about until 8.00 a.m. They were then asked to return to bed and two polyethylene intercaths were placed, one in each arm, in the antecubital space. Patients were requested to remain at bedrest for an additional hour and a half, and between 9.30 and 10.00 a.m. two plasma samples were drawn for measurement of AVP and osmolality. From 10.00 a.m. until noon, 5% hypertonic saline was infused by a constant infusion pump at the rate of 0.05 ml/kg/min. Blood was taken in the contralateral arm every

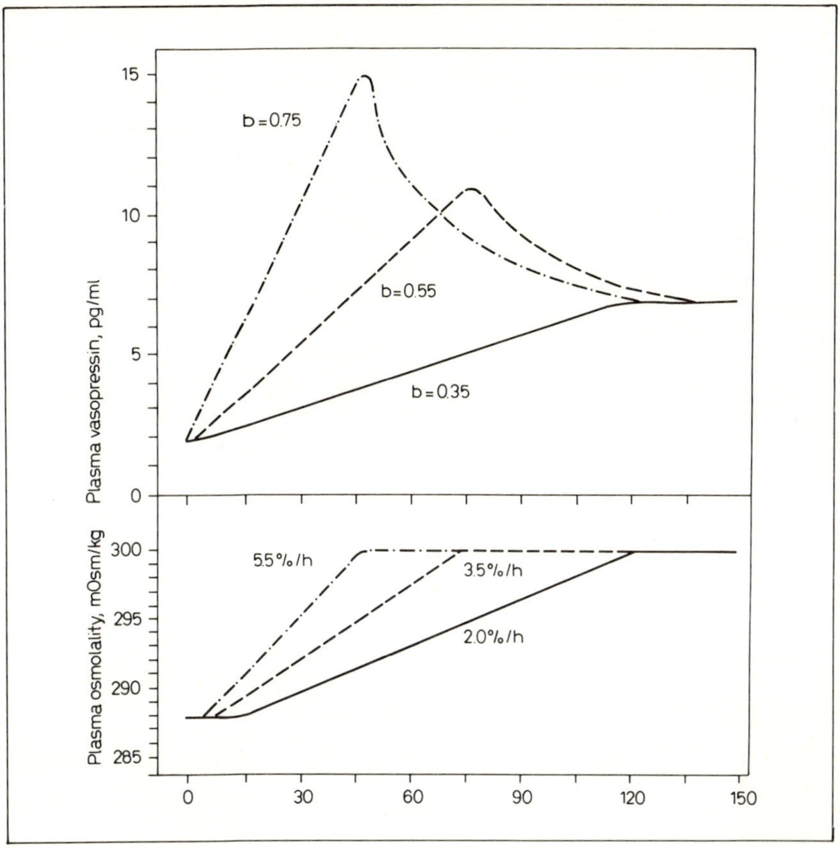

Fig. 1. Schematic representation of the effect of varying the rate of hypertonic saline infusion on the change in plasma vasopressin concentration in adults. At each rate (0.05, 0.1, 0.16 mg/kg/min) the total volume of saline infused was constant. The numbers juxtaposed to the lines in the lower panel indicate the overall rate of change in plasma osmolality, expressed as percent/per hour, during the infusions; in the upper panel, the numbers indicate the slope of the relationship between plasma vasopressin and osmolality during the same infusions. Courtesy of *Robertson and Athar* [unpublished observations].

15 min throughout the infusion and measured for AVP and osmolality. Because of the sensitivity of AVP secretion to acute changes in blood pressure, blood pressure was monitored every 5 min for 30 min before and throughout the procedure. If the mean blood pressure fell 10 mm from one determination to the next at any time during the procedure, the patient was excluded from the study.

The rate of infusion (0.05 ml/kg/min) has been shown to increase plasma osmolality at slightly below 2% per hour for at least 2 h [46]. Once the osmotic threshold for AVP secretion has been exceeded (usually greater than 280 mosm/kg) and plasma osmolality rises

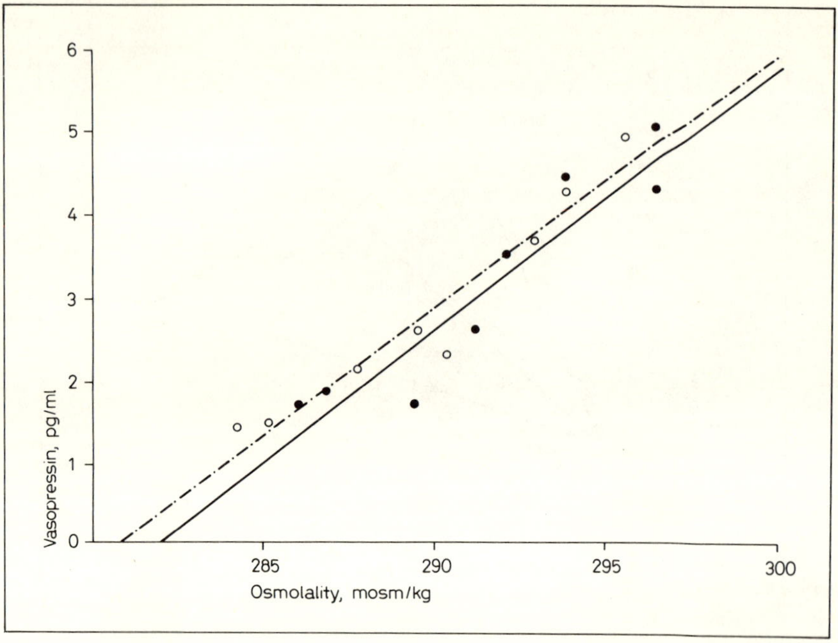
Vasopressin, pg/ml
0
1
2
3
4
5
6
285
290
295
300
Osmolality, mosm/kg

2

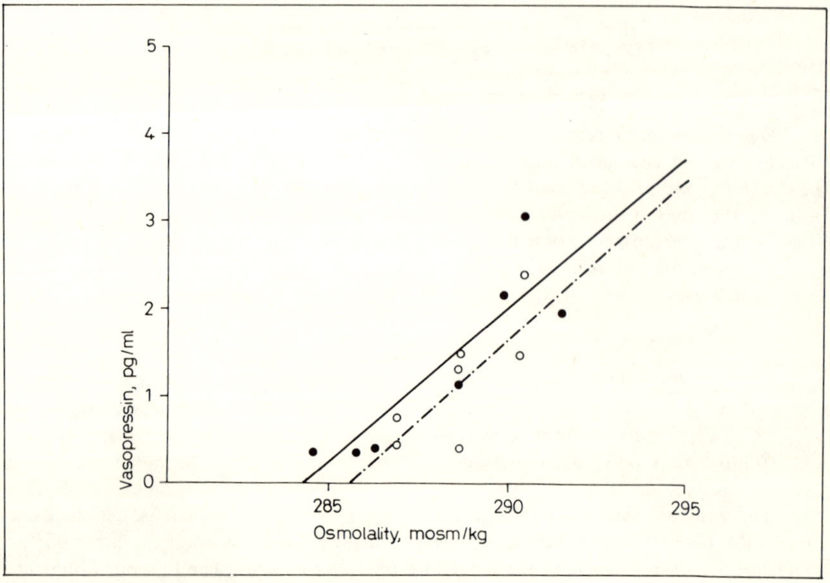
Vasopressin, pg/ml
0
1
2
3
4
5
285
290
295
Osmolality, mosm/kg

3

4

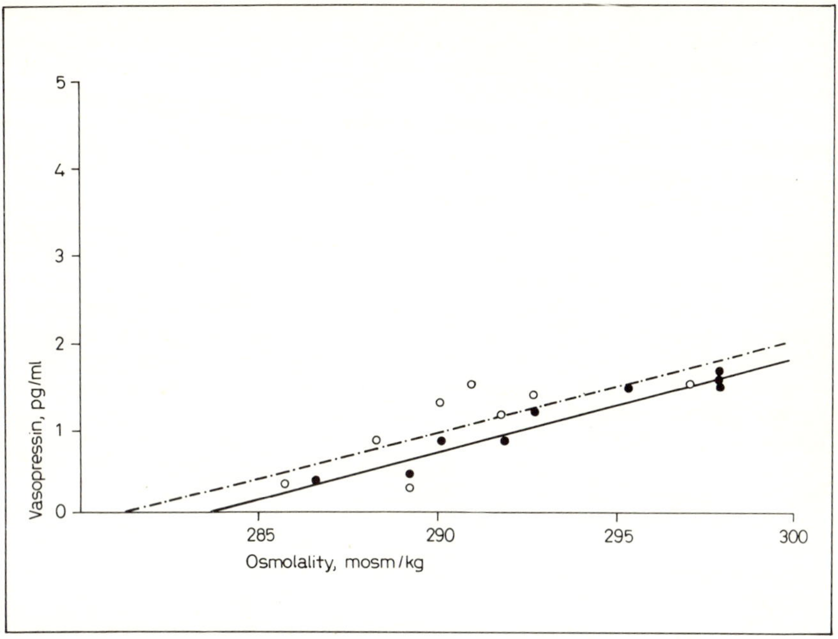

Fig. 2–4. Reproducibility of the AVP response to hypertonic saline in controls. Each subject was given the hypertonic saline infusion on two separate occasions at least 2 weeks apart.

at this rate, AVP has been shown to rise linearly correlated with the increase in plasma osmolality (fig. 1). This paradigm provides a quantitative measure of the major functional properties of the osmoregulatory apparatus. The slope of the regression line relates plasma AVP to osmolality and represents the sensitivity of the AVP response to increasing osmolality. The intercept at the abscissa represents the osmotic threshold for AVP secretion. The correlation coefficient (Pearson's r), represents the precision of the AVP response with the osmotic stimulus.

Two additional important features of data obtained by the hypertonic saline infusion are reproducibility and heritability. It has recently been shown that the threshold and sensitivity of the AVP response of 0.05 ml/kg/min of 5% saline is highly reproducible within a given individual over multiple infusions given even as long as 2 years apart [47]. The reproducibility of the AVP response to hypertonic saline within individuals makes the saline infusion paradigm ideal for the longitudinal study of psychiatric patients across shifts in clinical state and also the assessment of treatment effects of psychoactive agents on the osmoregulatory apparatus. Examples of this reproducibility in normal subjects are shown in figures 2–4. Heritability of the AVP response to hypertonic saline is indicated by the fact that the threshold and sensitivity of the AVP response to increasing osmolality are significantly correlated between monozygotic twins but not between dizygotic twins [47].

It should be noted that not only are the functional characteristics of the osmoreceptor in a healthy individual remarkably reproducible over time, but that they are remarkably resistant to modification by changes in blood volume and blood pressure. Thus, the 'set' of the osmoreceptor appears to change relatively little until blood volume decreases by a substantial amount (between 10–15% in man) [46], or until blood pressure is altered by a significant amount (± 10–15%) [48]. Thus, for accurate interpretation of hypertonic saline infusion data, it is not necessary to maintain patients on exquisitely controlled sodium and fluid regimens, but rather merely to ensure that intravascular volume, fluid and electrolyte balance, and blood pressure fall within a rather broad normal range.

3 drug-free subjects with bipolar affective illness were studied longitudinally with the hypertonic saline infusion paradigm during clinically established nonpsychotically depressed episodes and following the switch into mania. 3 psychotically depressed patients each received a single hypertonic saline infusion.

The plasma AVP response to hypertonic saline was monitored in patients who received either of two psychoactive agents with both antimanic and antidepressant effects, lithium and carbamazepine. 4 patients received hypertonic saline infusions during a placebo period and after 17–21 days of treatment with lithium carbonate at doses which ranged from 1,200 to 1,800 mg/day. Plasma lithium levels ranged from 0.9 to 1.2 mEq/l. 4 additional patients were studied longitudinally during teatment with placebo and following 17–30 days of treatment with carbamazepine in doses of 800–1,600 mg/day. Carbamazepine levels were maintained between 6 and 10 μl. Comparison of responses were made only if patients were studied longitudinally during the same clinical state.

Measurement of Osmolality and AVP in CSF and Plasma

The osmolality of CSF and plasma was determined by freezing-point depression (osmometer model 3L; Advanced Instruments, Needham Heights, Mass.). Unless otherwise indicated CSF and plasma were stored in glass vials at –70 °C for 4–12 weeks before extraction and assay.

AVP in CSF and plasma was assayed by the method of *Robertson* et al. [49]. Antiserum Glick-1 (Gl-1) used in these studies was provided by Dr. *Seymour Glick* [50]. This assay system effectively discriminates between arginine vasopressin, arginine vasotocin, and oxytocin. Serial dilutions of an acetone extract of human CSF and/or plasma produces a displacement curve that parallels that of purified synthetic AVP. Sensitivity of the assay is 0.4 pg/ml, intraassay variability 7%, and interassay variability 10%. All samples from one subject studied longitudinally were measured in the same assay.

Cognitive and Behavioral Effects of DDAVP Administration

DDAVP was selected as the AVP analog in the trial because compared with AVP, DDAVP shows enhanced antidiuretic, but markedly decreased pressor effects. Despite the loss of pressor activity, DDAVP was shown to exert cognitive effects analogous to that of naturally occurring AVP compounds in some, but not all, animal experimental paradigms [51].

DDAVP was administered intranasally in a calibrated catheter (rhinyle) in a solution containing 100 μg/ml. The placebo consisted of DDAVP vehicle (butanol in sterile water). DDAVP placebo was given for at least 7 days before active drug. Controls received 60 μg (20 μg t.i.d.) of active drug for 2–3 weeks; patients received 40–160 μg/day for 2–7 weeks.

During both the placebo and the active drug periods of the trial, patients had their weight and serum electrolytes recorded regularly, and their fluid intake was regulated daily to give an average of 1,400 ml/24 h. In this way the double-blind nature of the study was preserved while patients were protected against DDAVP-induced water retention.

Since previous findings indicate that the effects ofpeptide hormones may persist for some time after the last dose is administered [2], a randomized design was not used: all subjects started with a placebo period. However, all subjects practiced and familiarized themselves with tests of learning and memory, and placebo trials started only after no changes in the learning curve was evident. In addition, several unimpaired subjects were tested over a period of 8 weeks during which they received daily DDAVP placebo treatments; the experimenter administering the cognitive tests was blind to the fact that no active drug was used. Water intake was regulated and electrolytes monitored regularly just as in DDAVP-treated subjects.

The learning and memory tests used to assess the cognitive response after DDAVP treatment had each been previously tested in studies designed to investigate: (1) the cognitive response after drug treatment, including cholinergic and catecholaminergic agonists and antagonists [53, 54], and (2) the structure and extent of cognitive impairments in patients suffering from depression, mania, and various forms of dementia [55, 56]. Equivalent forms of these tests were used to provide measures of (1) serial learning and memory (subjects were repeatedly read at one word per 2 s, lists of 12 common unrelated words with serial-ordered recall after each presentation) [52]; (2) recall consistency or reliability using a prompted recall procedure (subjects were asked to remember as many words as possible from a list of 14 words read to them; only words not remembered on the previous trial were given after each attempted recall until the total list was monitored) [57], and (3) organization determinants that effect encoding of information and, thereby, learning and memory (a task that required subjects to recall 20 related words, made up of two sets of ten words each, drawn from common easily identified pretested categories, e.g., vegetables and flowers, presented in a random sequence for free recall 10 min later) [55].

4 severely depressed patients were included in the first study in which all of these parameters of cognitive function were extensively tested before and during DDAVP treatment; of these, 1 patient was severely psychotic. 3 additional patients with less severe depression are currently concluding DDAVP treatment and are also being tested. In addition to the depressed subjects, 6 cognitively unimpaired college student volunteers also participatet in the same regimen, receiving both placebo and DDAVP treatment and identical cognitive testing paradigms.

To test for DDAVP effects in ECT-induced retrograde amnesia a new procedure was devised in which patients were asked to listen to eight equivalent lists of eight related words each. The lists were controlled for their frequency in the language, their degree of semantic relatedness, and frequency of occurrences within a category [58]. A typical list was made up of eight vegetables (common and less common examples in a balanced arrangement), eight names of trees (similarly controlled), eight types of birds, etc. These eight lists were presented one at a time and the words within each list were presented at a 3 second rate. On actual ECT treatment days, the first list was presented 45 min prior to ECT, the second 5 min later, and so on, with the last presented 10 min prior to ECT treatment. Theoretically, the words given closest to the actual ECT treatment would be those most likely to be forgotten at the time of the subsequent recall trial, since ECT is typically reported to induce a pattern of retrograde amnesia. This recall trial was given 5 h after the ECT treatment, when patients were alert. On recall, subjects were first given the name of the category from which items were drawn and then asked to remember the words.

2 subjects were studied during the course of 6 weeks of biweekly unilateral ECT treatments. Each subject was given te cognitive paradigm on three occasions. The first administration was for baseline purposes prior to any actual ECT treatments. DDAVP placebo 0.1 ml t.i.d. in each nostril was started 2 days after the first sign of ECT-induced lifting of mood. After 5 days on DDAVP placebo the next test was given. After the test the patients were given active DDAVP, 0.1 ml in each nostril t.i.d. 60 μg/day for 5 days at the end of which the cognitive tests were repeated.

Results

CSF AVP Studies

No significant sex differences in CSF AVP levels and no correlation with age was found in either the patient or the control group. Two bipolar euthymic patients were subjected to 3 h of induced hyperactivity in the early morning just prior to the LP as previously described [59]. In these subjects, CSF AVP showed no significant change, suggesting that sleep loss and/or hyperactivity do not appear to effect CSF AVP.

CSF AVP was significantly lower in drug-free nonpsychotic unipolar and bipolar depressed patients (0.88 ± 0.07 pg/ml) compared with controls (1.21 ± 0.14 pg/ml) ($p < 0.04$) (table I, fig. 5). CSF AVP levels were extremely high in the 2 psychotically depressed patients (4.1 and 3.1 pg/ml), higher than in any of the 40 patients. There was no significant difference in CSF osmolality among any of the patient subgroups.

CSF AVP was also significantly lower in nonpsychotic bipolar depressed patients (0.78 ± 0.07 pg/ml) compared with patients in the manic or hypomanic state (1.38 ± 0.16 pg/ml) ($p < 0.01$) (table I, fig. 6); while CSF osmolality did not differ significantly between depressed and manic patients. There was a trend for the CSF AVP to be lower in the nonpsychotic bipolar depressed group compared with controls ($p < 0.1$) (table I, fig. 7).

CSF AVP was also studied in 9 patients during a drug-free euthymic period. The mean of 1.03 ± 0.13 pg/ml was not significantly different from the mean levels seen in the overall depressed group or that of controls (table I, fig. 5).

Hypertonic Saline Infusion

4 drug-free bipolar patients were given hypertonic saline infusions during the depressed and manic phases of their illness, respectively. There was no significant differences in the mean threshold for AVP secretion between the depressed (280.5 mosm/kg) and manic phase (281 mosm/kg). The precision during both depression and mania ($r = 0.99$ and 0.94, respectively) was excellent. The sensitivity of the AVP response to hypertonic saline was greater during the manic phase than during the depressive phase in each of the three subjects studied longitudinally. Thus, for the first subject the sensitivity in depression was 0.26 pg/ml/mOsm/kg compared to a sensitivity of 0.40 pg/ml/mOsm/kg in mania. For subject 2 the sensitivity in depression was 0.11 pg/ml/mOsm/kg and 0.23 pg/ml/mOsm/kg in mania; for subject three the sensitivity in depression was 0.39 pg/ml/mOsm/kg and rose to 0.89 pg/ml/mOsm/kg in mania.

3 psychotically depressed patients were given hypertonic saline infusions. They all showed a normal threshold (between 280 and 284 mosm/kg), and excellent precision ($r > 0.90$). Interestingly, the sensitivity of the response in each of these subjects was greater than 1.0 pg/ml/mosm/kg.

Table I. CSF AVP in patient subgroups and controls

Groups	Patients	CSF AVP
Controls	32	1.21 ± 0.14
		↓
		$p < 0.04$
		↑
Nonpsychotic depressed; UP = 12; BP = 9	21	0.88 ± 0.07
Psychotic depressed; UP = 1; BP = 1	2	3.6 ± 0.50
BP nonpsychotic depressed	9	0.78 ± 0.07
		↓
		$p < 0.01$
		↑
Hypomanic and manic[1]	8	1.38 ± 0.16
Euthymic; UP = 2; BP = 7	9	1.03 ± 0.13

Data expressed as mean ± SEM; significance established by Student's t test.
[1] Independent sample of patients in the manic state.

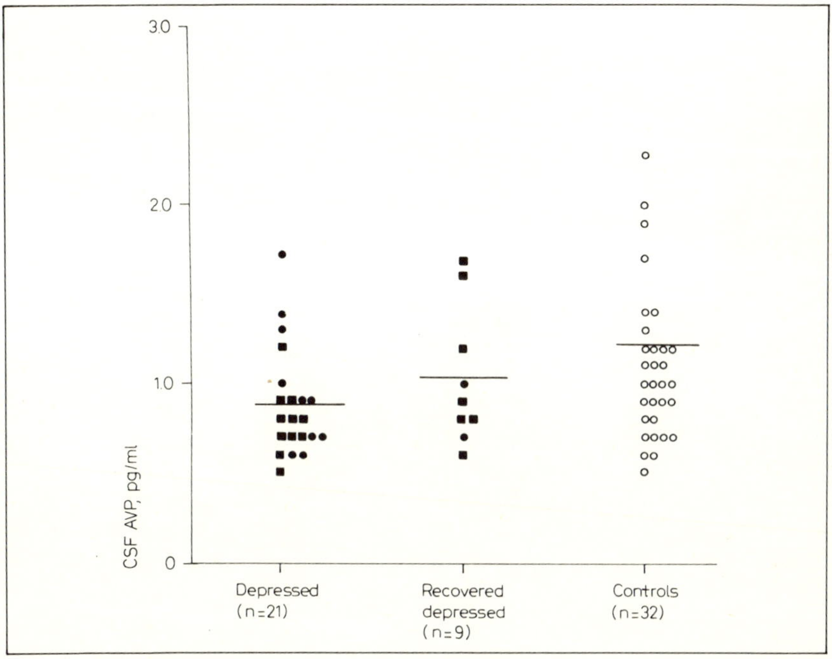

Fig. 5. CSF AVP in drug-free depressed and revovered patients and controls. Depressed vs. controls: $p < 0.04$. ■ = Bipolar, ● = unipolar.

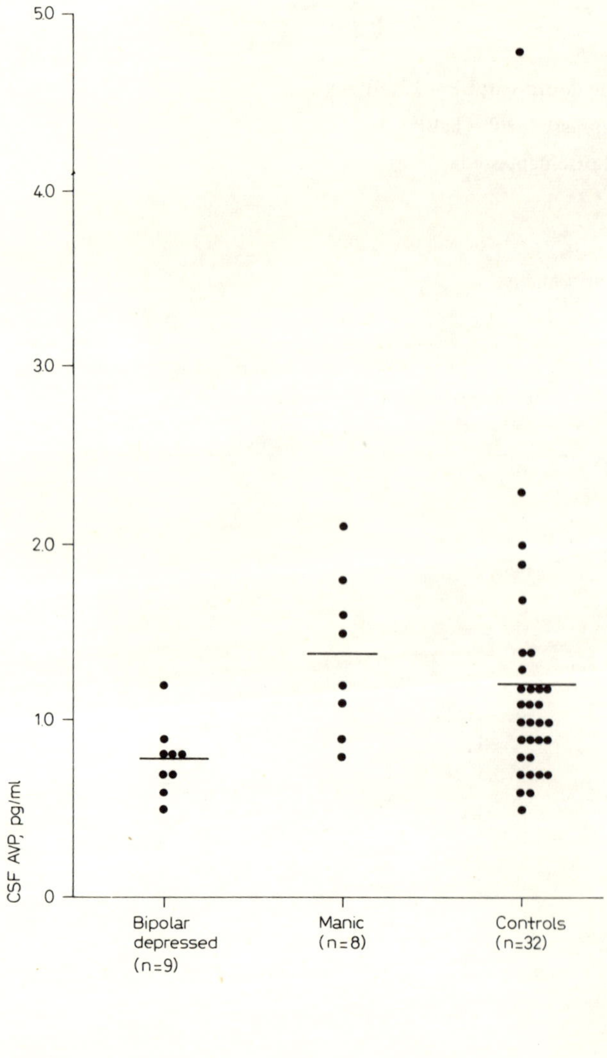

Fig. 6. CSF AVP in drug-free bipolar patients and controls. Depressed vs. manic: $p < 0.01$.

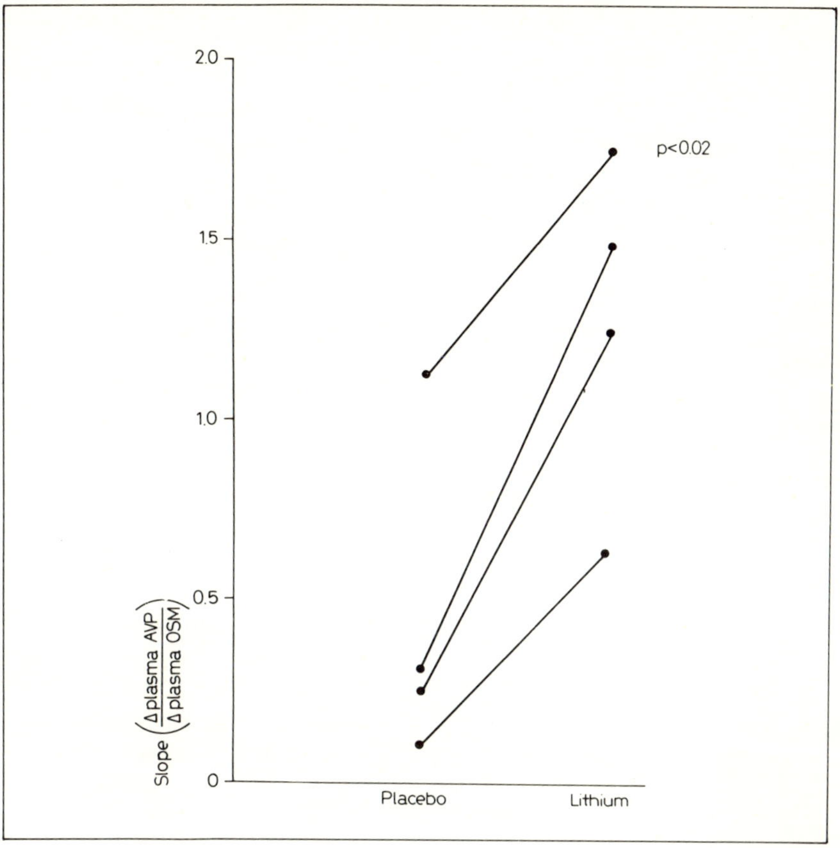

Fig. 7. Sensitivity of the plasma AVP response to hypertonic saline in patients with affective illness tudied longitudinally during a placebo period and following lithium. Each data point represents the slope of a regression line generated by at least eight separate plasma AVP and osmolality (OSM) determinations.

4 subjects (2 in the depressed, 2 in the manic phase of their illness) were studied before and during chronic lithium treatment. Lithium did not significantly influence the threshold of AVP secretion. Precision was excellent during both placebo ($r = 0.95$) and lithium ($r = 0.96$) treated periods. Lithium significantly increased the sensitivity of AVP secretion in this group from 0.4 pg/ml/mosm/kg during placebo to 1.27 pg/ml/mosm/kg ($p < 0.01$, paired t) (fig. 7) The relatively high mean sensitivity in the placebo group (0.42 pg/ml/mosm/kg) is due to 2 patients in the sample who were manic; lithium significantly in-

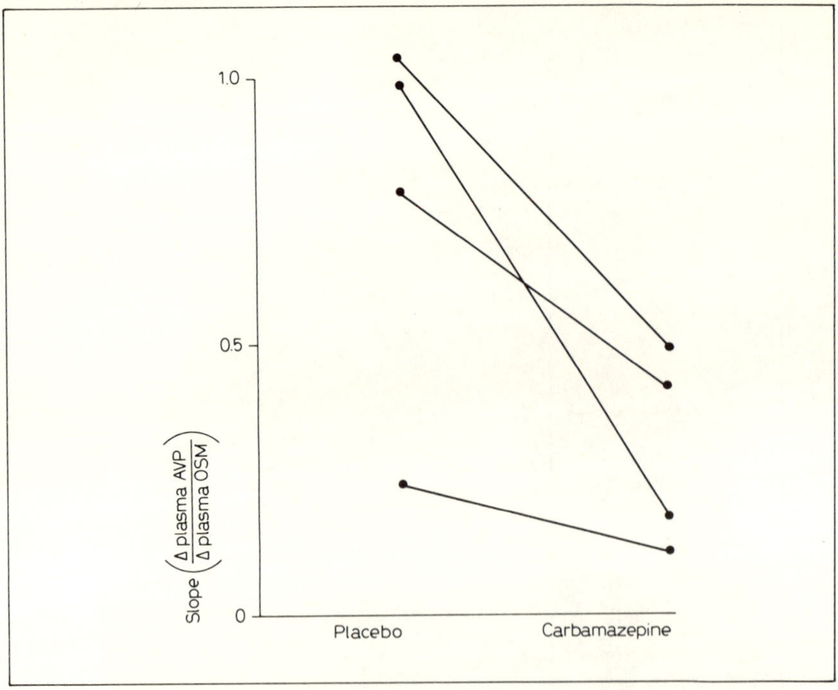

Fig. 8. Plasma AVP response to hypertonic saline in patients with affective illness studied longitudinally during a placebo period and carbamazepine. Each data point represents the slope of a regression line.

creased the sensitivity in these patients despite the persistence of the manic symptoms.

4 subjects were also studied before and after carbamazepine treatment. No significant effect of carbamazepine on threshold was found; precision was excellent during both the placebo and carbamazepine treatment period ($r = 0.94$ and 0.96, respectively). Carbamazepine lowered the sensitivity from 0.042 to 0.15 pg/ml/mosm/kg ($p < 0.1$) (fig. 8). The high value of the sensitivity mean (0.42 pg/ml/mosm/kg) during the placebo period is due to 2 of 4 depressed patients who were psychotic; 14–17 days of carbamazepine treatment lowered the sensitivity of the response despite persistence of the psychosis.

Cognitive Response to Vasopressin Analog Administration

4 depressed patients with impaired cognition related to their disordered mood were treated with DDAVP for at least 3 weeks after completion of a peri-

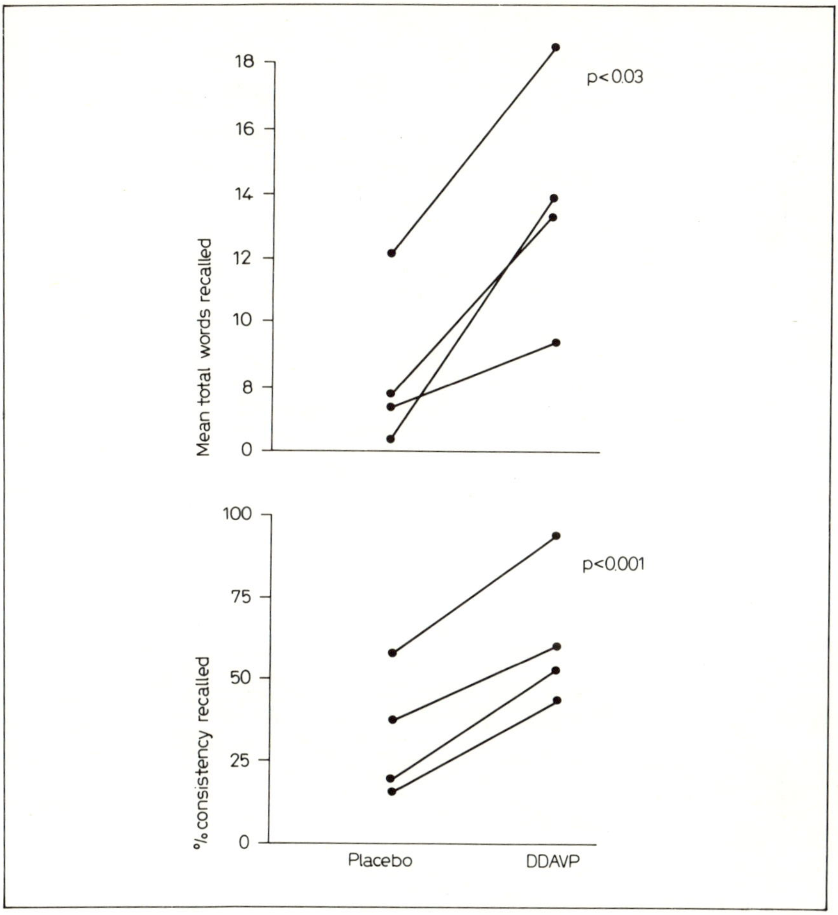

Fig. 9. Cognitive effects of DDAVP in depressed patients. Each data point represents the mean value of at least three tests per subject of mean total words recalled, and percent consistency recalled.

od of baseline assessments of learning, memory, and mood. 2 of these patients showed an amelioration of depressed mood (see below) while the overall group demonstrated significant cognitive enhancements beginning 2 days after DDAVP treatment, and continuing for some time (2 weeks) following DDAVP treatment. The statistically significant increase in learning and memory function was evident on measures of: (1) increased recall of information on the prompted recall task, with an average increase of $23 \pm 0.07\%$ ($p < 0.01$, Anova with repeated

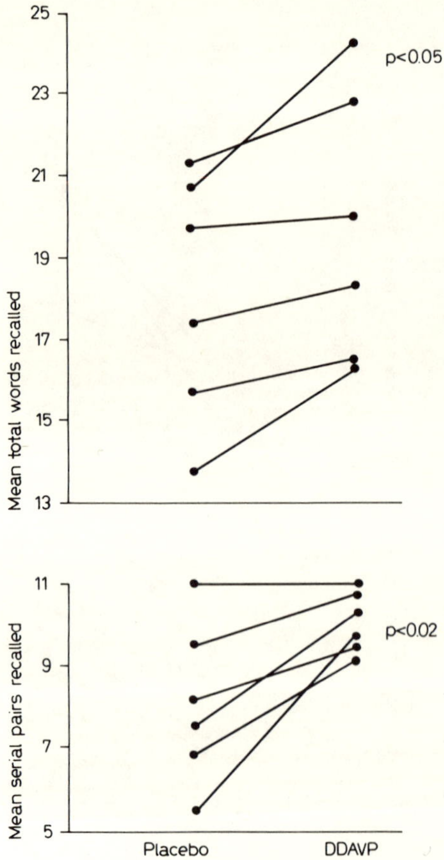

Fig. 10. Cognitive effects of DDAVP in cognitively unimpaired controls. Each data point represents the mean value of at least three tests per subject of mean total words recalled, and mean serial pairs recalled.

measures); (2) a related marked increase in the percent consistency (or reliability) of recall of previously remembered words from a mean of 46 ± 0.08 to 0.63 ± 0.06% ($p < 0.001$) (fig. 9); and (3) an increase in free recall of semantically related words ($p < 0.03$) (mean total words recalled; fig. 9), along with an increase in the organization of remembered words, as measured by the tendency to cluster or remember together related words on recall ($p < 0.01$). Serial learning, a task that proved to be particularly difficult for these patients (few of the patients reaching criterion) was not improved after DDAVP treatment. Patients had not returned to their baseline learning-memory performance until 4 weeks after active drug treatment had ceased. These changes in learning and memory were not significantly correlated with changes in the severity of clinical depression, although the patients did spontaneously report feeling more activated, and expressed that they felt more able to think clearly while being treated with DDAVP.

6 young normal cognitively unimpaired subjects demonstrated statistically significant increases in learning and memory after treatment with DDAVP, in contrast to placebo treatment. Unimpaired subjects demonstrated a DDAVP-related increase in recall of semantically related words ($p < 0.05$) (mean total words recalled; fig. 10), in serial learning ($p < 0.02$) (mean serial pairs recalled), and in prompted free recall (most marked at the end of a period of 2 weeks of DDAVP treatment; $p < 0.01$). There was, however, no DDAVP-related changes in the consistency of recall of previously remembered words (information) stored in long-term memory. This was probably due to a ceiling effect; even during placebo conditions subjects reliably recalled previously remembered words (consistency of recall of $91 \pm 0.04\%$). Matched cognitively unimpaired college student volunteers treated with DDAVP placebo three times a day for a month, and tested twice weekly at the same time of day, failed to demonstrate any systematic change in these measures of learning and memory.

Following ECT, but without DDAVP treatment, both patients demonstrated a retrograde amnesia: a decrease in recall as a function of time of presentation, prior to ECT treatment. Recall of words presented at least 25 min prior to ECT were not disrupted but recall of the last four lists, presented 20 and fewer minutes prior to ECT was reduced by two-thirds that of baseline values (tested by an exact probability statistic; $p < 0.01$). The retrograde amnesia effect was substantially reversed following DDAVP treatment, as indicated by a threefold increase in recall of items presented closest in time (20 min or less) to ECT administration (tested by an exact probability statistic; $p < 0.01$). Despite the amelioration of retrograde amnesia, the patients overall mental status did not seem improved.

Behavioral Responses to Vasopressin Analog Administration

2 of 4 severely depressed patients showed significant amelioration of depressed affect. The most dramatic clinical response occurred in a severely psychotically depressed patient who had delusions and loud auditory hallucinations almost continuously for 8 months before treatment. She was given 60–80 μg/day of DDAVP for 7 weeks. 48 h after DDAVP was started the depression and psychosis improved substantially. After 3 weeks of intermittent good and bad days, she showed significant reduction in the weekly average nurse's ratings [44] of depression and psychosis (including almost complete cessation of auditory hallu-

cinations) that persisted for the remainder of the trial ($p < 0.001$). There was a gradual but ultimately profound relapse into psychotic depression within 3 weeks after discontinuation of DDAVP treatment. Cognitive performance rose sharply 48 h after treatment began and remained at this level throughout treatment and for at least 3 weeks after discontinuation of DDAVP despite fluctuation in clinical state. Repeat cognitive testing 6 weeks after cessation of treatment showed a return of cognitive function to baseline levels.

The second patient who also showed significant amelioration of depressed affect was continuously depressed for 5 months prior to treatment, received 80 μg of DDAVP daily for 4 weeks. She felt much less depressed on the third day of treatment, and for the second and third weeks of the trial depression ratings were significantly reduced ($p < 0.001$). However, this response was transient and she relapsed while still on active treatment. Cognitive performance remained enhanced throughout the trial despite the relapse and declined 2 weeks after DDAVP was discontinued.

A third patient with a severe psychotic depression of 12 months duration showed no mood response to DDAVP but had a robust cognitive response on a dose of 60–80 μg/day. During the 3 days when the dose was raised above 120 μg/day cognitive functioning deteriorated considerably. 10 days after DDAVP was stopped, cognitive function declined to baseline levels. A fourth patient who had regular cycles of depression and hypomania showed no statistically significant cognitive or affective response to DDAVP, though the hypomanic phase of her illness appeared substantially lengthened. All patients had a brief hyponatremia.

We are just completing a study of 3 additional depressed patients who were cognitively less impaired and less depressed than the original 4 patients tested. Preliminary analysis shows that 2 of these 3 patients experienced a significant cognitive response to DDAVP, but none showed significant amelioration of the affective component of their depression. Thus, DDAVP influenced depression ratings in only 2 of 7 patients, and had no significant effect on the mean depression ratings in the overall group of subjects.

Discussion

CSF AVP

CSF AVP is significantly lower in a group of drug-free unipolar and bipolar depressed patients compared with controls. Among bipolar patients, CSF AVP was significantly lower in nonpsychotically depressed patients compared with patients studied in the manic or hypomanic phase. Psychotic depression was associated with a marked elevation of CSF AVP. These findings represent the first report of significant differences in CSF AVP levels between psychiatric patients and control subjects or within patient subgroups. They are not a reflection of the influence of hydration status on levels of AVP since CSF sodium and osmolality do not differ significantly between patients and controls or among any of the patient subgroups. Blood pressure changes, nausea, or prior drug ingestion can also be excluded as variables influencing these observations. One previous report noted no significant differences between CSF AVP in drug-free unipolar depressed patients compared with controls [60]. Several of the subjects in the above study were psychotically depressed, so that a direct comparison with our results is not possible. There is no previous report comparing CSF AVP levels in drug-free bipolar depressed and manic patients.

The hypothalamic neurons that secrete AVP into the CSF seem to be anatomically separate from those that secrete into the blood, suggestive of the presence of a separate, specialized, central AVP secretory system [61, 62]. The presence of a discrete central AVP secretory apparatus is strongly supported by studies of patients with idiopathic diabetes insipidus who show elevated CSF AVP levels despite subnormal levels in plasma [61, 62]. Thus, the neurosecretory pathway to the CSF in these individuals seems spared the irreversible degeneration of the supraoptico-hypophyseal tract that is noted in patients with surgical or other causes of diabetes insipidus. Elevated CSF AVP in patients with central diabetes insipidus suggests that CSF AVP is not a reflection of passive diffusion of hormone from plasma, and thus supports the existence of a special central neurosecretory pathway to the CSF. It is of interest that this central AVP pathway seems to have little, if anything, to do with systemic water balance, a role generally attributed to AVP; patients with idiopathic diabetes insipidus despite elevated levels of CSF AVP, continue to exhibit polydipsia, polyuria, and a deficiency of plasma AVP. Although the role of this central secretory system has not been elucidated, it seems possible that the CSF AVP pathway evolved to facilitate AVPs modulation of several aspects of CNS function, such as cognition and central fluid and electrolyte balance. In support of this idea is the finding that intraventricular administration of anti-AVP antibody disrupts memory function in previously intact experimental animals [14].

Whether changes in CSF AVP levels are related to syndromal aspects of affective illness is unclear. One could speculate that diminished memory function in nonpsychotic depressed patients is related to relatively diminished AVP levels in the CSF. Cognitive enhancement in depressed patients following DDAVP administration supports but does not confirm this idea. The significance of the relatively increased CSF AVP in mania and psychotic depression is also unclear; whether these findings relate to the accelerated thinking in mania or to the ruminative, perseverative quality of thought in psychotic depression, is also only a matter of speculation at this time. However, a recent report noted the precipitation of a hypomanic episode in a patient after treatment with AVP for senile dementia [63].

Several possible mechanisms could contribute to changes in CSF AVP levels. Abnormalities in synthesis or metabolism could each contribute to our findings. Changes in CSF AVP could also result from alterations in the storage, release, or actions of any of the neuropeptides or neurotransmitters that modulate the set-points for the physiological variables that govern CSF AVP secretion. For example, if there were an increase in the 'set' or sensitivity of the hypothalamic osmoreceptor, CSF AVP levels would rise even if osmolality remained constant. Since osmolality is a major physiologic regulator of CSF AVP, we have decided to elucidate the characteristics of the hypothalamic osmoreceptor in patients with affective illness. This we have done by measuring the plasma AVP response to a sustained intravenous infusion of hypertonic saline.

Plasma AVP Response to Hypertonic Saline Infusion

Drug-free bipolar patients show significantly lower sensitivity of the plasma AVP response to hypertonic saline during the depressive (nonpsychotic) compared with the manic or hypomanic phase of their illness. Psychotically depressed patients showed the highest sensitivity of any subgroup tested. The data represent the first demonstration of differences in osmoregulation between subgroups of patients with psychiatric illness. Lithium and carbamaezapine, two drugs with both antimanic and antidepressant properties, significantly influence the sensitivity of the AVP response to hypertonic saline. These findings are the first demonstration of a drug effect in humans on the plasma AVP response to hypertonic saline. None of these differences in sensitivity can be attributed to differences in the factors known to shift sensitivity in humans, such as hydration status, blood pressure, or emesis.

Because of the difficulties involved in measuring the small quantities of AVP present in biological fluids, only two previous studies have been published that report plasma AVP levels in psychiatric patients. In the first of these, *Raskind* et al. [64] noted that compared with controls, baseline plasma AVP levels were significantly higher in acutely psychotic (schizophrenic and depressed) patients than in severely anxious nonpsychotic patients. These data are compatible with our report of increased CSF AVP and sensitivity of plasma AVP response to hypertonic saline in drug-free psychotically depressed patients. *Raskind* et al. [65] also reported that ECT causes a marked transient rise in plasma AVP levels.

Increased sensitivity of the plasma AVP response to an osmotic stimulus in mania compared with depression means that at a given osmolality the AVP level in plasma, and presumably CSF will be higher in mania than in depression. This relative increase in sensitivity during mania compared with depression could contribute to some or all of the increased CSF AVP noted in manic compared with depressed patients. The relatively exaggerated sensitivity (slope greater than 1 pg/ml/mosm/kg vs. normal mean slope of 0.34 pg/ml/mosm/kg) was noted in

3 of 3 psychotically depressed patients; the increased sensitivity of the osmoreceptor in psychotic depression could contribute significant to the very high CSF AVP levels seen in these patients.

Changes in osmoreceptor sensitivity could result from alterations in the functional activity of any of the neuromodulators at any synapse or extrasynaptic site between the osmoreceptor and the neurohypophysis. Changes in the permeability of the osmoreceptor membrane in different clinical states could also influence the sensitivity of the osmoreceptor, since at a given osmolality the osmoreceptor volume, and hence AVP secretion, would differ because of this circumstance. The situation is complicated by the fact that certain evidence suggests that the osmoreceptor may operate in a bimodal fashion, i.e. the osmoreceptor suppresses as well as stimulates vasopressin release. This hypthesis, which is consistent with the known behavior of certain other regulatory functions of the hypothalamus, would explain the phenomenon observed in some patients with pathological osmoreceptor function who are unable to suppress AVP even when overhydrated and hyponatremic [24]. Such an hypothesis is also consistent with recent electrophysiological studies that have shown that the hypothalamus contains two populations of osmosensitive neurons, one of which is stimulated and the other suppressed by hypertonicity [66].

Effects of Pharmacologic Agents on Osmoregulation

There are no previous studies examining the effects of pharmacologic agents on the AVP response to osmotic stimuli in man. To date, preliminary work from our group shows that in 4 normal controls naloxone (15 mg s.c.) has no effect on the AVP response to hypertonic saline; therefore opiates may not play a physiologic role in osmoregulation [21] despite considerable evidence that acute opiate administration significantly increases plasma AVP levels [22]. We found that two psychopharmacologic agents with both antidepressant and antimanic properties, lithium [67] and carbamazepine [40], significantly affect the sensitivity of the AVP response to hypertonic saline in depressed patients. These drugs have been previously reported to alter AVP function in experimental animals and to influence water clearance in humans [42]. Thus, lithium was shown to be an AVP receptor antagonist at the renal level [68] and often induces nephrogenic diabetes insipidus in patients who take the drug chronically [41]. Carbamazepine, on the other hand, is an AVP functional agonist, and though there are no data in the literature to suggest a mechanism, carbamazepine is often used to treat or augment the treatment of partial central diabetes insipidus [42]. Thus, lithium and carbamazepine have opposite effects on overall AVP function in experimental animals and in humans.

In patients with affective illness, lithium and carbamazepine also seem to have opposite effects on AVP secretion to a hypertonic stimulus. Thus, lithium, the AVP receptor antagonist, increases the sensitivity of the AVP response to hypertonic saline; carbamazepine, the functional agonist, causes a decrease in the sensitivity and seems to act as an AVP receptor agonist. Lower levels of AVP are needed to promote antidiuresis and maintain a given plasma osmolality when hypertonic saline is administered during carbamazepine treatment compared with the drug-free state. Although the effects of lithium and carbamazepine on the AVP response are in the opposite direction, the two drugs can be seen to exert analogous effects on the system; each promotes changes in the plasma AVP response in the direction opposite to its effects on the renal AVP receptor. Speculatively, lithium and carbamazepine may exert therapeutic effects in bipolar affective illness by initiating compensatory changes in AVP secretion in the direction opposite to their effects on a putative central AVP receptor. Possibly, these drugs function by circumscribing the range of central AVP action similar to their established effects on other regulated systems [69]. This formulation represents a somewhat different concept of psychotropic drug action, in which drug efficacy is linked to the stabilization of the functional activity of a central peptidergic neuromodulatory system.

Cognitive and Behavioral Responses to DDAVP Treatment

DDAVP significantly enhanced the formation, encoding, and organization of long-term trace events in memory in the initial group of 4 depressed patients. DDAVP also induced a similar pattern of cognitive enhancement in a group of cognitively unimpaired normal volunteers. Previous reports show that AVP can improve memory function in humans with amnesia [63] and senile dementia [70].

Naturally occurring analogs of AVP are known to have different biological activities, and synthetic analogs of AVP are available in which some of the biological effects of the naturally occurring peptide have been altered or abolished. Compared with AVP, DDAVP has a prolonged half-line, increased antidiuretic, decreased pressor, and absent corticotropin (and presumably beta-endorphin) releasing activity [46]. No reports exist on whether or not DDAVP possesses AVP's reported ability to alter central biogenic amine metabolism [16] or other biological or physiological processes of relevance to primary affective disorder. However, some of the behavioral effects of DDAVP have been studied in experimental animals. It has been reported that DDAVP is markedly less effective than AVP in delaying extinction to aversive stimuli [51]. *De Wied* [51] attributes this lack of activity to substitution of *L*-arg-8 by *D*-arg in DDAVP. On the other hand, it has been shown that DDAVP is slightly more active than AVP in reversing puromycin-induced amnesia [71]. These differences cannot be explained at present, but they may be related to the test system used. It is of interest that a naturally occurring cleavage product of oxytocin, pro-Leu-Gly-NH_2 has been reported to also reverse puromycin-induced amnesia [71] and to ameliorate depressive symptomatology in humans [72].

The design of the DDAVP study raises the question that repeated administration of the cognitive tests during the placebo period contributed to the enhancements noted during DDAVP treatment. It seems unlikely that practice effects significantly influenced the results. Normal controls to whom these tests were repeatedly administered do not show any systematic practice effects. In this study each subject was carefully familiarized with the procedures prior to testing. They also showed a relatively stable baseline over at least five testing sessions and none of the subjects showed a greater than 10% difference between the first baseline test and the mean of the remaining baseline values. Moreover, in each of the patients in whom cognitive enhancements occurred, performance declined considerably, at variable periods of time, after discontinuation of the drug.

The structure of the cognitive changes following DDAVP treatment in depressed patients and controls is analogous to that seen when patients recover from depression. Specifically, the formation of long-term (relatively permanent) trace events in memory and their effect on coding and organization are altered more than those processes that are either attentional or measures of immediate or short-term memory. The patterns of enhancement resemble those that have been noted after amphetamine administration [73] rather than those that were reported after cholingergic stimulation [53]. Quantitatively, the improvements noted after DDAVP administration are considerably greater than those that have been noted following either recovery from depression [52] or after amphetamine administration [73].

Only 2 of 7 depressed patients responded to DDAVP with a significant decrease in depression ratings. It should be mentioned, however, that the one patient who showed the most robust response relapsed after the drug was stopped, despite the continued administration of DDAVP placebo. In addition, all 7 patients who participated in the trial had been refractory to all other standard treatments including lithium, tricyclics, and, in some cases, ECT. Nevertheless, this preliminary experience with DDAVP as an antidepressant is discouraging. Perhaps AVP itself which, in contrast to DDAVP, has corticotropin and endorphin releasing properties, will prove to be a better antidepressant agent.

Conclusion

AVP is a representative central peptide with a multiplicity of CNS effects of possible relevance to psychiatric illness. The strategies and data presented here illustrate the feasibility of studying central peptide function in a coordinated and comprehensive fashion.

Differences in CSF AVP between subgroups of patients with affective illness may relate to alterations in the sensitivity or set of the hypothalamic osmoreceptors that govern the secretion of AVP into both CSF and plasma. The possible linkage between CSF AVP levels and the sensitivity of osmoregulation in subgroups of patients amplifies the importance of elucidating the factors that influence the sensitivity of the osmoreceptor in humans. Fortunately, osmoregulation in humans is amenable to direct study via hypertonic saline infusion conducted during drug-free states and following treatment with neurotransmitter agonists and antagonists and peptide hormone administration.

Whether alterations of CSF and plasma AVP levels or function are related to the basic pathophysiology or certain aspects of the symptomatology of affective illness is not known. From a diagnostic point of view, a very high CSF AVP level and sensitivity to osmotic stimuli may be characteristic of psychotic depression though further work is needed both with depressed and psychotic patients sampled from a variety of diagnostic catagories to evaluate this possibility.

The DDAVP trial represents an example in which an hypothesis and clinical 'work-up' of a central peptide system led to a specific therapeutic intervention. The fact that DDAVP enhances specific aspects of information processing which are most altered in depressed states is of considerable interest. Whether vasopressin analogs will prove clinically useful in some treatment-resistant patients is yet to be determined.

Summary

AVP is a hypothalamic peptide secreted directly into the systemic circulation and widely distributed in brain via axon transport and CSF. AVP seems to be involved in a variety of areas that show dysfunction in affective illness, such as memory, biological rhythms, REM sleep, pain perception, neuroendocrine regulation, fluid and electrolyte balance, and the response to psychoactive agents. Three separate paradigms were used to evaluate AVP function in patients with affective illness: (1) measurement of AVP in CSF; (2) assessment of the threshold, precision, and sensitivity of the plasma AVP response to intravenous hypertonic saline infusion, and (3) clinical evaluation of long-term AVP analog treatment.

We found that CSF AVP is significantly lower in drug-free depressed (nonpsychotic) patients than in controls. In bipolar patients, CSF AVP is also significantly lower in the depressed (nonpsychotic) compared with the manic state. Psychosis in depressed patients is associated with a marked elevation of CSF AVP. Evaluation of the plasma AVP response to hypertonic saline suggests that the differences in CSF AVP levels among subgroups of patients with affective illness may be accounted for by specific alterations in the sensitivity or 'set' of hypothalamic osmoreceptors that govern AVP secretion into both CSF and plasma.

Depressed patients and cognitively unimpaired normal volunteer subjects responded to DDAVP, a synthetic analog of AVP, with significant functional improvement in the specific components of cognition known to be impaired in the depressed state. DDAVP also significantly reversed ECT-induced retrograde amnesia. While DDAVP did not show a significant effect on the mean depression ratings for the overall group of severely depressed patients, there were 2 patients in whom periods on DDAVP administration, in contrast to placebo periods, were associated with clear improvement in mood and/or in psychotic features.

References

1 Weindl, A.; Sofroniew, M.V.; Schinko, I.: The distribution of vasopressin, oxytocin, neurophysin, somatostatin, and luteinizing hormone releasing hormone producing neurons; in Bargmann, Oksche, Polenov, Scharrer, VIIth Int. Symp. on Neurosecretion, Leningrad 1978, p. 308.

2 Barker, J.L.: Physiological role of peptides in the nervous system; in Gainer, Peptides in neurobiology, p. 295 (Plenum Press, New York 1977).

3 Barker, J.L.; Gainer, H.: Peptide regulation of bursting pacemaker activity in a molluscan neurosecretory cell. Science *184:* 1371 (1974).

4 Wied, D. de; Bohus, B.: Long-term and short-term effects on retention of a conditioned avoidance response in rats by treatment with long acting pitressin and a MSH. Nature, Lond. *212:* 1484 (1966).

5 Stromgren, L.S.: The influence of depression on memory. Acta psychiat. scand. *56:* 109 (1977).

6 Kupfer, D.: REM latency: a psychobiological marker for primary depressive disease. Biol. Psychiat. *11:* 159 (1977).

7 Vogel, G.W.; Vogel, F.; McAbee, R.S.; Thurmond, A.J.: Improvement of depression by REM sleep deprivation. New findings and a theory. Archs gen. Psychiat. *37:* 247 (1980).

8 Byck, R.: Lithium ion, endorphins, and mood. Lancet *ii:* 721 (1975).

9 Wehr, T.A.; Wirz-Justice, A.; Duncan, W.; Gillin, J.C.; Goodwin, F.K.: Phase advance of the circadian sleep-wake cycle as an antidepressant. Science *206:* 710 (1979).

10 Murphy, D.L.; Goodwin, F.K.; Bunney, W.E., Jr.: Electrolyte changes in the affective disorders: problems of specificity and significance; in Katz, Shield, Recent advances in the psychobiology of the depressive illness, p. 59 (U.S. Government Printing Office, Washington 1972).

11 Carroll, B.J.; Curtis, G.S.; Mendels, J.: Neuroendocrine regulation in depression. II. Discrimination of depressed from non-depressed patients. Archs gen. Psychiat. *33:* 1051 (1976).

12 Sachar, E.J.; Frantz, H.G.; Altman, W.: Blunted growth hormone response to amphetamine infusion in depression. Am. J. Psychiat. *130:* 134 (1973).

13 Wied, D. de; Bohus, B.; Wimersma Greidanus, T.J.B. van: Memory deficit in rats with hereditary diabetes insipidus. Brain Res. *85:* 152 (1975).

14 Wied, D. de: Behavioral effects of intraventricularly administered vasopressin and vasopressin fragments. Life Sci. *19:* 685 (1976).

15 Roche, K.E.; Lehsner, A.I.: ACTH and vasopressin treatments immediately after a defeat increase future submissiveness in male mice. Science *204:* 1343 (1979).

16 Tanaka, M.; Kloet, E.R. de; Wied, D. de; Versteeg, O.: Arginine vasopressin affects catecholamine metabolism in specific brain nuclei. Life Sci. *20:* 1799 (1977).

17 Urban, I.; Wied, D. de: Neuropeptide's effects on paradoxical sleep and theta rhythms in rats. Pharmacol. Biochem. Behav. *8:* 51 (1978).

18 Longo, V.G.; Loizzo, A.: Evaluation of the central effects of ergot alkaloids by means of electroencephalography. Pharmacology *16:* suppl. 1, p. 189 (1978).

19 Van Ree, J.M.; Wied, D. de: Prolyl-leucyl-glycinomide facilitates neurophine dependence. Life Sci. *19:* 1331 (1977).

20 Wied, D. de; Gispen, W.H.: Behavioral effects of peptides; in Gainer, Peptides in neurobiology, p. 397 (Plenum Press, New York 1977).

21 Gold, P.W.; Extein, I.; Pickar, D.; Goodwin, F.K.; Robertson, G.L.: The effect of acute

and chronic methadone on the plasma arginine vasopressin response to hypertonic saline in depressed patients (in preparation).

22 Weitzman, R.E.; Fisher, D.W.; Minisek, S.; Ling, N.; Guilleman, F.: Beta-endorphin stimulates secretion of arginine vasopressin in vivo. Endocrinology *101:* 1643 (1977).

23 Mata, M.; Gainer, H.; Klee, W.: Effect of dehydration on the endogenous opiate content of the rat neuro-intermediate lobe. Life Sci. *21:* 1159 (1979).

24 Robertson, G.L.: The physiopathology of ADH secretion; in Tolis, Clinical neuroendocrinology: a pathophysiological approach, p. 247 (Raven Press, New York 1979).

25 Zimmerman, E.; Robinson, A.G.: Hypothalamic neurons secreting vasopressin and neurophysin. Kidney int. *10:* 12 (1976).

26 Sofroniew, M.V.; Weindl, A.: Projections from the parvocellular vasopressin and neurophysin containing neurons of the suprachiasmatic nucleus. Am. J. Anat. *153:* 391 (1978).

27 Steton, M.H.; Watson-Whitmyre, M.: Nucleus suprachiasmaticus. The biological clock in the hamster. Science *191:* 197 (1976).

28 Ruzak, B.; Zucker, I.: Biological rhythms and animal behavior. A. Rev. Psychol. *26:* 137 (1975).

29 Kleeman, C.R.; Rubini, M.E.; Lambdin, E.; Ebstein, G.H.: Studies on alcohol diuresis. II. The evaluation of alcohol as an inhibitor glycinamide (MIF-1) on dopaminergic neurons. Pharmacol. Biochem. Behav., suppl. 1, p. 125 (1955).

30 Rundgren, M.; Leksell, G.; Lishajko, F.; Andersson, B.: Deuterium-induced extinction of ADH release in response to intracerebroventricular infusions of hypertonic NaCl and angiotensin. Acta physiol. scand. *100:* 45 (1979).

31 Coppen, A.; Shaw, D.M.; Malleson, A.: Changes in 5-hydroxytryptophan metabolism in depression. Br. J. Psychiat. *111:* 105 (1965).

32 Gold, P.W.; Robertson, G.L.: Central paptide regulation of fluids and electrolyte balance: psychiatric implications; in Alexander, Fluid and electrolyte disturbances in psychiatry (Spectrum Press, New York, in press).

33 Raichle, M.D.; Grubb, R.L., Jr.: Regulation of brain water permeability by centrally released vasopressin. Brain Res. *143:* 191 (1978).

34 Rudman, D.; Chawla, R.K.: Antidiuretic peptide in mammalian choroid plexus. Am. J. Physiol. *230:* 50 (1976).

35 Martin, J.B.; Reichlin, S.; Brown, G.: Regulation of growth hormone secretion and its disorder; in Clinical neuroendocrinology, p. 147 (Davis, Philadelphia 1977).

36 Kwaan, H.C.; Bartelstone, J.H.: Corticotropin release following injections of minute doses of arginine vasopressin in the third ventricle of the dog. Endocrinology *65:* 982 (1959).

37 Lutz-Bucher, B.; Koch, B.; Mialhe, C.: Comparative in vitro studies on corticotropin releasing activity of vasopressin and hypothalamic median eminence extract. Neuroendocrinology *23:* 181 (1977).

38 Yates, F.E.; Maran, J.W.: Stimulation and inhibition of adrenocorticotropin release; in Handbook of physiology, section 7: Endocrinology, vol. IV, part 2, p. 367 (American Physiology Society, Washington 1979).

39 Goodwin, F.K.; Dunner, D.L.; Gershon, E.S.: Effect of *L*-dopa treatment on brain serotonin metabolism in depressed patients. Life Sci. *16:* 751 (1971).

40 Ballenger, J.C.; Post, R.M.; Bunney, W.E., Jr.: Carbamazepine (Tegretol) in manic-depressive illness: a new treatment. Am. J. Psychiat. *137:* 7 (1980).

41 Singer, I.; Rotenberg, D.; Puschett, J.: Lithium-induced nephrogenic diabetes insipidus: in vivo and in vitro studies. J. clin. Invest. *51:* 1081 (1972).

42 Moses, A.M.; Miller, M.; Streeten, D.H.P.: Pathophysiologic and pharmacologic alterations in the release and action of ADH. Metabolism *25:* 697 (1976).

43 Spitzer, R.L.; Endicott, J.; Robins, E.: Research diagnostic criteria: rationale and reliability. Archs gen. Psychiat. *35:* 773 (1978).

44 Kotin, J.; Goodwin, F.K.: Depression during mania: clinical observations and theoretical implications. Am. J. Psychiat. *129:* 679 (1972).

45 Luerssen, J.G.; Robertson, G.L.: Unpublished observations (1980).

46 Robertson, G.L.; Athar, S.: The interaction of blood osmolality and volume in regulating plasma vasopressin in man. J. clin. Endocr. Metab. *42:* 613 (1976).

47 Zerbe, R.L.; Miller, J.Z.; Robertson, G.L.: Reproducibility and heritability of vasopressin osmoregulation in humans. Annu. Meet. Endocrine Society, Washington 1980.

48 Robertson, G.L.; Kinney, R.A.; Nelson, A.E.: The effect of isoproterenol on vasopressin secretion. Proc. 56th Annu. Meet. Endocrine Society, Atlanta. Endocrinology *94:* suppl., p. 217 (1974).

49 Robertson, G.L.; Maller, E.A.; Athar, S.; Sinha, T.: Development and clinical application of a new method for the radioimmunoassay of arginine vasopressin in human plasma. J. clin. Invest. *52:* 2340 (1973).

50 Ommaya, S.N.; Kagan, A.; Glick, S.M.: Radioimmunoassay of vasopressin: application to unextracted human urine. J. clin. Endocr. Metab. *33:* 739 (1977).

51 Wied, D. de: Peptides and behavior. Life Sci. *20:* 195 (1977).

52 Henry, G.M.; Weingartner, H.; Murphy, D.L.: Influence on affective states and psychoactive drugs on verbal learning and memory. Am. J. Psychiat. *130:* 966 (1973).

53 Sitaram, N.; Weingartner, H.; Caine, E.D.; Gillin, G.C.: Choline: selective enhancement of serial learning and encoding of imagery words in man. Life Sci. *22:* 1555 (1978).

54 Sitaram, N.; Weingartner, H.; Gillin, G.C.: Human serial learning: enhancement with arecholine and choline and impairment with scopolamine. Science *201:* 274 (1978).

55 Weingartner, H.; Cohen, R.; Murphy, D.L.: Structure of cognitive deficits in depression and mania. Archs gen. Psychiat. (in press).

56 Caine, E.D.; Ebert, M.H.; Weingartner, H.: Cognitive function in primary affective disorder. Neurology *27:* 1087 (1977).

57 Buschke, H.: Selective reminding for analysis of learning and memory. J. verb. Lrng. verb. Behav. *12:* 543 (1973).

58 Battig, W.F.; Montague, S.: Category norms for verbal items in 56 categories: a replication and extension of the Connecticut Category Norms. J. exp. Psychol. Monogr. *80:* 1 (1969).

59 Post, R.M.; Kotin, J.; Goodwin, F.K.; Gordon, E.K.: Psychomotor activity and cerebrospinal fluid amine metabolites in affective illness. Am. J. Psychiat. *130:* 67 (1973).

60 Gjerris, A.; Fahrenkrug, J.; Hammer, M.; Schaffalitzky del Muckadell, O.B.; Hojholm, S.; Vendsborg, P.; Rafaelsen, O.J.: Peptides in the cerebrospinal fluid of manic-melancholic and schizophrenic patients; in Abstr. 12th CINP Congress, p. 155 (Pergamon Press, Göteborg 1980).

61 Luerrsen, J.G.; Robertson, G.L.: Cerebrospinal fluid vasopressin and vasotocin in health and disease; in Neurobiology of cerebrospinal fluid (Plenum Press, New York, in press).

62 Luerrsen, J.G.; Shelton, G.L.; Robertson, G.L': Evidence for separate origin of plasma and cerebrospinal fluid vasopressin. Clin. Res. *25:* 14A (1977).

63 Audibert, A.; Benghier, P.: Influence of vasopressin on learning and memory. Lancet *i:* 41 (1978).

64 Raskind, M.A.; Weitzman, R.E.; Orenstein, H.; Fisher, D.A.; Courtney, N.: Is antidiuretic hormone elevated in psychosis. A pilot study. Biol. Psychiat. *13:* 385 (1978).

65 Raskind, M.A.; Orenstein, H.; Weitzman, R.E.: Vasopressin in depression. Lancet *i:* 164 (1979).
66 Vincent, J.D.; Arnauld, E.; Bioulac, B.: Activity of osmosensitive single cells in the hypothalamus of the behaving monkey during drinking. Brain Res. *44:* 371 (1978).
67 Goodwin, F.K.; Murphy, D.L.; Bunney, W.E., Jr.: Lithium response in unipolar vs. bipolar depression. Am. J. Psychiat. *129:* 44 (1972).
68 Dousa, T.; Hechter, O.: The effect of NaCl and LiCl on vasopressin sensitive adenyl cyclase. Life Sci. *9:* 765 (1970).
69 Mandel, A.: Neurological mechanisms of presynaptic adaptation and their organization; in Mandel, Neurobiological mechanism of adaptation and behavior, p. 1 (Raven Press, New York 1975).
70 Oliveros, J.C.; Jandol, M.K.: Timsit-Berthier, M.; Remy, R.; Benghezala, A.; Audibert, A.; Moeglen, J.M.: Vasopressin in amnesia. Lancet *i:* 42 (1978).
71 Walter, R.; Hoffman, P.L.; Flexner, J.B.; Flexus, L.B.: Neurohypophyseal hormones, analogs, and fragments: their effect on puromycin-induced amnesia. Proc. natn. Acad. Sci. USA *72:* 4180 (1978).
72 Kastin, A.J.; Miller, M.C.; Ferrel, L.: General activity in intact and hypophysectomized rats after administration of melanocyte-stimulating hormone (MSH) melatonin and Pro-Leu-Gly-NH2. Physiol. Behav. *10:* 399 (1973).
73 Reus, V.I.; Silberman, E.K.; Post, R.M.; Weingartner, H.: *D*-Amphetamine: affects on memory in a depressed population. Biol. Psychiat. *14:* 345 (1979).

P.W. Gold, MD, Clinical Psychobiology Branch, National Institute of Mental Health, 9000 Rockville Pike, Building 10, Room 4S239, Bethesda, MD 20205 (USA)

Adv. biol. Psychiat., vol. 7, pp. 71–84 (Karger, Basel 1981)

Receptor Function in Depression

William E. Bunney, Jr., Blynn Garland
Biological Psychiatry Branch, National Institute of Mental Health, Bethesda, Md., USA

Introduction

This paper will review some strategies for attempting to study receptor function in man, including data specifically relevant to manic-depressive illness, and will propose a hypothesis concerning the possible role of altered receptor function in affective illness. In addition, animal studies will be reviewed concerning the effects of lithium carbonate, one of the most effective therapeutic agents in manic-depressive illness, on receptor function.

There are a number of methodological difficulties involved in attempting to study neurotransmitter receptor function in man. Most available agonists and antagonists have multiple sites of action. Some of the specific drugs which have been particularly useful in animals, such as 6-hydroxydopamine or 5, 6-dihydroxytryptamine, are not available for use in man because of their apparent irreversible neurotoxicity. Another difficulty in studying receptor function in the central nervous system in man is that often only indirect response measures are available. Blood pressure, pulse, salivation, pupillary response and temperature are some examples of measurable responses. Sleep parameters provide a possible quantitative and sensitive measure of neuronal receptor change. Physical activity is another measure utilized in animals along with stereotypy. There is a literature suggesting that neuroendocrine responses may be useful [*Meltzer* et al., 1978; *Prange* et al., 1979; *Pandey* et al., 1977]. Dopamine agonists, for example, are associated with a decreased release of prolactin and luteinizing hormone and in increased growth hormone responses [*Meltzer* et al., 1974; *Sachar,* 1973; *Donoso* et al., 1971; *Kamberi* et al., 1971]. Also, gross behavioral ratings of psychosis, mania and depression may be shown to reflect alterations in receptor site sensitivity. It is necessary in man to look at final common reactions such as sleep and activity responses. When studying whole body indices of receptor function it is important to emphasize that many intervening variables may be responsible for the final response, with receptor function being only one such variable. Nevertheless, a number of interesting strategies are currently being systematically experimentally tested in man in an attempt to obtain indirect measures of receptor function. Both pharmacological and tissue models have been developed for attempting to estimate receptor function in man [*Bunney,* 1981]. Tissue models involve the utilization of lymphocytes to study peripheral beta-receptors, platelets to study peripheral alpha-receptors, and brain tissue to evaluate a number of possible receptors.

Receptor Strategies

Beta-Adrenergic Receptors

By assaying melatonin's secretion and by using differential doses of propranolol (a beta-blocker), it may be possible to arrive at an index of beta-adrenergic receptor function. Melatonin's synthesis and secretion are initiated and maintained by the release of norepinephrine (NE) at the neuroendocrine synapse with the consequent stimulation of the pineal's beta-adrenergic receptor. These receptors have been studied in the rat and show a diurnal rhythm of responsiveness. The human pineal is also under beta-adrenergic control. Measurement of plasma melatonin may provide an in vivo technique for assaying changes in beta-adrenergic receptor sensitivity in humans. Alterations in beta-adrenergic sensitivity might be expected in a disease associated with increased sympathetic activity such as mania. In a preliminary study of 4 manic patients, *Lewy* et al. [1979] showed that these patients, when compared to controls, have increased plasma melatonin as measured by a negative ionization gas chromatographic-mass spectrometry technique. In other studies [*Jimerson* et al., 1977; *Lynch* et al., 1978] urinary melatonin rhythms were analyzed. It was demonstrated that urinary melatonin excretion by normals peaked during the hours of darkness [*Lynch* et al., 1975]. *Jimerson* et al. [1977] compared 8 healthy controls with 6 depressed patients and found that both groups had similar diurnal rhythms with increased melatonin secretion during the night. In addition, the two groups showed no difference in rhythmic secretions following sleep deprivation. Depending on the dose, propranolol decreases or blocks nocturnal melatonin secretion. By utilizing different dosages of propranolol administered prior to sleep, it is conceivable that one could develop dose-response curves in varying populations of patients or normals that might give an index of beta-adrenergic receptor sensitivity. A fixed dose of propranolol, which partially decreased melatonin secretion, might be utilized in a given individual in a longitudinal study during both manic and depressive phases.

Tissue Models

It has been proposed that beta-adrenergic receptor function on human leukocytes could be used as peripheral tissue models for the study of changes in beta-adrenergic receptor sensitivity in brain. Two methods have been used for the evaluation of beta-adrenergic activity. The first method involves the measurement of leukocyte adenylate cyclase and its responsiveness to NE and the beta-adrenergic agonist, isoproterenol. Using this methodology, *Pandey* et al. [1979] reported decreased beta-adrenergic activity in 9 manic and 21 depressed patients compared to 26 controls. The second method involves the measurement of specific binding of beta-adrenergic agonist, *L*-^{3}H-dihydroalprenolol (^{3}H-DHA), to lymphocytes. Using this method *Extein* et al. [1979] reported decreased beta-adrenergic activity in 9 manics and 13 depressed patients as compared to 16 controls and 10 euthymic patients. However, this group was unable to replicate the findings using adenylate cyclase measurements. The authors speculate that the decrease may either be an index of change in beta-adrenergic receptors or may simply reflect homeostatic regulation of peripheral beta-adrenergic receptor response to stress-induced increases in circulating catecholamines [*Extein* et al., 1979].

Chronic Treatment with Antidepressants

A number of recent investigators, including *Sulser* [this volume], have documented that most of the successful treatments of affective illness including tricyclic antidepressants, monoamine oxidase inhibitors and electroconvulsive therapy have all been associated with a down-regulation of beta-adrenergic receptors in animals and in man. There is also a growing body of evidence that the down-regulation of the beta-receptor associated with chronic antidepressant medication may be mediated in response to presynaptic stimulation of the alphareceptor.

Imipramine-Serotonin Receptors in Human Platelets

Recent work has shown that a high affinity binding site for ^{3}H-imipramine exists in human platelets and that their properties are apparently similar to those high affinity binding sites for ^{3}H-imipramine in rat brain [*Raisman* et al., 1979]. In addition, high affinity binding sites for ^{3}H-imipramine have also been reported in human brain [*Rehavi* et al., 1980]. These sites were found to be saturable, reversible, and inhibited by pharmacologically active tricyclic antidepressants. Other psychoactive drugs were ineffective in inhibiting binding to human brain sites. A clinical study of *Briley* et al. [1979, 1980] measured high affinity binding of ^{3}H-imipramine to platelet membranes in 16 untreated depressed patients and 21 age-matched controls of the same sex. The binding in the depressed group was significantly lower than in the controls, although there was no change in affinity. The results suggest that the decreased binding in depression may reflect these changes in the brain. It has also been shown that there is a relationship between the binding site for ^{3}H-imipramine and the presynaptic reuptake site for serotonin since the relative potencies of the tricyclic antidepressants as serotonin reuptake blockers and as inhibitors of ^{3}H-imipramine are quite high [*Paul* et al., 1980].

Acetylcholine Receptors

REM Sleep and Arecholine (a Muscarinic Cholinergic Agonist)

Using the induction of REM sleep as a whole body index of neurotransmitter function, it may be possible to study acetylcholine receptor mechanisms in man. Arecholine is a muscarinic cholinergic agent agonist and if given to man after the first REM period it can decrease the usual duration from the first to the second REM period. The time from the first to the second REM period is selected due to the variability in the time from sleep onset to the first REM period in depressed patients. In a series of studies, *Sitaram and Gillin* [in press] have shown that arecholine produces a statistically significant shorter time interval from the end of the first REM period to the beginning of the second in patients with affective illness as compared to normals. Affectively ill patients in remission (those patients off drugs for up to 4 months) show a shortened REM latency between the first and second periods, which is similar to that observed following administration of arecholine [*Sitaram and Gillin* in press]. When two different dosages of arecholine are administered, a rough dose-response curve can be obtained with a clear shift to the left in those patients who are affectively ill. The data suggest that there is a possible supersensitive muscarinic cholinergic receptor. In another series of studies *Sitaram* et al. [1979] administered scopolamine, a muscarinic cholinergic agonist, to normal volunteers on three consecutive mornings. It was hypothesized that supersensitivity would develop by the evening hours, as the central effects of the drug have been demonstrated to last only 4–6 h [*Ketchum* et al., 1973]. In contrast to scopolamine given at night when there was a prolongation of REM latency [*Sagales* et al., 1968, 1975], there was a decrease in REM latency. When arecholine was given at night following the daily morning dose of scopolamine, there was a potentiation of the effect with a further decrease in REM latency. This could be interpreted as a production of supersensitive muscarinic acetylcholine receptors following scopolamine treatment in the morning, which then amplified the effect of arecholine given at night. Thus, the use of muscarinic agonists and antagonists and the whole body measure of REM latency may provide an index of muscarinic cholinergic receptor sensitivity.

Relation of Pupil Size to Estimate Muscarinic Cholinergic Receptor Sensitivity

In an effort to obtain another whole body index of receptor function in man the human pupil has been proposed as a model system for studying both adrenergic and muscarinic cholinergic drug response. The neural input to the sphincter muscles are cholinergic. Evidence from *Sitaram and Gillin* [in press] provides pupillometric evidence in support of supersensitivity, which is compatible with the observations derived from the arecholine studies. Thus two bodies of data are compatible with the possibility of increased muscarinic receptor sensitivity in depressed patients.

Alpha Receptors

Pupillometry

Pupillometry may also be used as a system for the study of adrenergic systems in man as there is an alpha-adrenergic input to the dilator muscles. Ocular pharmacology has a large literature describing dose-dependent effects of adrenergic drugs, as well as cholinergic drugs, on the size of the pupil [*Ishikawa* et al., 1976; *Mapstone,* 1976; *Smith and Smith,* 1978].

Antidepressants

A recent study of *Maggi* et al. [1980] administered various antidepressants to rats and measured the amount of inhibition to the alpha-2-adrenergic agonist, clonidine. The data suggests that antidepressants are weak alpha-2-adrenergic receptor blocking agents, but that some are more potent on the alpha-1-adrenergic receptor WB4101, indicating that any reducing potential of some antidepressants may not be mediated by the blockade of presynaptic autoreceptors [*Maggi* et al., 1980]. In a review article on presynaptic NE regulation during depression, *Cohen* et al. [1980] suggest that the development of supersensitive alpha-2-adrenergic receptors may lead to affective illness in vulnerable individuals. From preliminary work, *Cohen* et al. [1980] suggest that in response to antidepressants there is a functional presynaptic hyposensitivity that develops over a 2- to 3-week period, possibly through the release of NE into the synapse.

Dopamine Receptors

Apomorphine and Hypothermia

Apomorphine, a direct dopamine agonist, produces quantifiable hypothermia in man and might provide an index of dopamine receptor function. Apomorphine's hypothermic effect involves a central dopaminergic component in animals. *Ary* et al. [*1977*] have demonstrated hypothermia in animals after intraventricular administration of either apomorphine or dopamine. *Quock and Gale* [1974] and more recently *Cox and Lee* [1977] suggest that the brain area mediating apomorphine hypothermia is the preoptic region of the hypothalamus, since direct iontophoretic application of apomorphine and dopamine onto these cells produces hypothermia, while pretreatment with haloperidol and pimozide blocks this response. Furthermore, *Cox and Lee* [1979] suggested with behavioral studies that apomorphine appears to be inducing hypothermia by lowering the hypothermic temperature set-point and not through peripheral action.

Cutler et al. [1979] quantified changes in temperature in man following apomorphine and demonstrated that hypothermic effects of apomorphine but not other response measures such as nausea and shivering were significantly greater at 9.00 p.m. than at 9.00 a.m. It is possible that the apparent changes in sensitivity to apomorphine in man in the evening and in the morning may reflect differences in receptor sensitivity. This methodology might prove useful in the study of conditions in which abnormalities of brain dopamine receptors have been proposed, such as manic-depressive illness, schizophrenia, tardive dyskinesia and parkinsonism [*Bunney* et al., 1977b; 1979; *Post* et al., 1978a; *Pandey* et al., 1977; *Smith* et al., 1977]. In a recent animal study, *Schelkunov* [1980] demonstrated that apomorphine in piribedil-induced hypothermia in mice could be blocked by dopaminergic agents such as neuroleptics, but were unaffected by anticholinergic agents such as scopolamine or atropine, which is compatible with the finding that the stimulation of dopamine receptors plays a role in hypothermic response.

Manic-Depressive Illness (Hypothesized Supersensitive Catecholamine Receptors Associated with the Onset of Mania)

Bipolar manic-depressive patients often change from a retarded, nonverbal, dozing depression into hyperactive mania. We have hypothesized [*Post* et al., 1976] that the switch into mania may be associated with the development of supersensitive catecholamine receptors and that lithium carbonate, which is the compound most effective in blocking the behavioral switch into mania, may block the development of supersensitive dopamine receptors. Animal data has accumulated which is supportive of a hypothesis that lithium blocks the development of supersensitive dopamine receptors. This data, along with the effects of lithium on the alterations in receptor sensitivity in other neurotransmitter systems, will be reviewed.

Behavioral observations during the onset of mania are compatible with a marked increase in released catecholamine neurotransmitters. This behavior is similar to and, at times, indistinguishable from behavior associated with the intravenous infusion of a large dose of amphetamine. The switch into mania is associated with a sudden increase in motor and verbal activity, intrusiveness, increased interest in sex and absence of sleep. Similarly, behavioral changes associated with relatively large doses of intravenous amphetamine include increased motor and verbal activity, increased interest in sex, and decreased sleep. Amphetamine increases release and decreases reuptake of catecholamines, particularly dopamine and NE, thus producing indirect catecholamine agonist actions. While amphetamine rarely produces typical mania [*Gerner* et al., 1976], it does provide a drug model that produces both functional increases in brain catecholamines and many behavioral effects similar to those seen at the onset of mania [*Bunney* et al., 1977a].

Measurements of catecholamines and indoleamines and their metabolites have been completed in the urine and CSF over the past decade during both manic and depressive phases of the illness. In reviewing this data it is difficult to explain the onset of mania on the basis of a marked release of indoleamines or catecholamines as reflected by the metabolites, 3-methoxy-4-hydroxyphenylglycol in the urine and CSF, and homovanillic acid and 5-hydroxyindoleacetic acid in the CSF [*Bunney* et al., 1977b]. The most that can be said is that some authors have reported a moderate increase in these amines in patients who switch from depression into mania. However, the levels in mania do not appear to exceed those noted in normal controls. Recent data [*Post* et al., 1978b] does suggest an increase in CSF NE during mania. Thus, if one is to hypothesize that the onset of mania is associated with a marked increase in functional catecholamine activity, additional mechanisms such as receptor sensitivity changes should be considered in a primary or adjunctive role. Although pharmacological data has accumulated in the past suggesting a role for NE in mania, recent data have also documented a role for dopamine in the manic process. Briefly, administration of the catecholamine precursor, *L*-dopa, has been reported to be associated with hypomania in bipolar patients [*Murphy* et al., 1971] while alpha-methyl-*p*-tyrosine (AMPT), a specific inhibitor of the rate-limiting step in the synthesis of NE and dopamine, has been shown to decrease manic episodes [*Brodie* et al., 1971]. However, fusaric acid, which is a dopamine-beta-hydroxylase inhibitor, was not effective in decreasing manic symptomatology [*Sack and Goodwin,* 1974]. This suggests that dopamine rather than NE may be involved in mania, since blockade of the synthesis of dopamine and NE is effective in mania, while blocking synthesis following the formation of dopamine and prior to the formation of NE is ineffective in decreasing mania. Finally, pimozide, the most specific neuroleptic blocker of dopamine receptors, has been shown to be effective in decreasing mania [*Post* et al., in press].

One of the most consistently striking characteristics of the onset of mania is the loss of sleep during the day prior to and on the day of the onset of the manic episode. Sleep changes following the discontinuation of AMPT provide an interesting pharmacological model, with sudden insomnia and, in some patients, the onset of manic-like symptoms [*Bunney* et al., 1977a].

Since lithium blocks the onset of mania, which we hypothesize may be associated with supersensitive receptors, studies have been undertaken to evaluate the effect of lithium carbonate on blocking the development of dopamine receptor supersensitivity, alpha- and beta-receptor supersensitivity, and serotonin receptor supersensitivity.

Dopamine Receptor Supersensitivity

Chlorpromazine and other neuroleptics, when given chronically to rats and then discontinued, can produce dopamine receptor supersensitivity as tested by observing stereotypy, motor behavior or by measuring ^{3}H-haloperidol receptor binding following discontinuation of a neuroleptic. Lithium alone has had no effect on apomorphine-induced stereotypy, locomotor behavior, or on ^{3}H-haloperidol receptor binding in the striatum. Chronic treatment with neuroleptics, tyrosine hydroxylase inhibitors, reserpine, or interruption of dopaminergic transmission by lesions of the nigrostriatal dopamine pathway, have all been shown to produce an increased sensitivity to behavioral effects of drugs that stimulate dopamine receptors. Supersensitivity induced by chronic administration of neuroleptics [*Muller and Seeman,* 1978; *Burt* et al., 1977], as well as nigrostriatal lesions [*Creese* et al., 1977], have been found to be accompanied by an increase in striatal dopamine receptor sites. It has been proposed that this apparent proliferation of the receptors following chronic blockade of dopamine transmission may be causally related to behavioral supersensitivity [*Burt* et al., 1977; *Creese* et al., 1977].

Pert et al. [1978], after long-term treatment of rats with haloperidol, produced a statistically significant increased sensitivity to locomotor and stereotypy effects of the dopamine agonist apomorphine. However, the rats treated with lithium 1 week before and during haloperidol administration failed to develop the behavioral sensitivity response to apomorphine, as measured by locomotor and stereotypic behavior [*Pert* et al., 1978]. Scatchard analysis of this data revealed that the increase in receptor binding in the haloperidol-treated group was due to an increase in the number of binding sites and not alteration in the affinities of the existing binding sites [*Pert* et al., 1978]. It was also shown that lithium did not increase the rate of haloperidol excretion or decrease the concentration of haloperidol at the dopamine receptors in the rat striatum. Behavioral findings similar to the above were demonstrated in rats pretreated with AMPT. Following discontinuation of AMPT these rats developed evidence of behavioral supersensitivity to apomorphine that was blocked by pretreatment with lithium [*Pert* et al., unpublished observation].

Electrophysiological data utilizing single unit recording demonstrated evidence of supersensitive presynaptic receptors in the zona compacta following chronic haloperidol, as demonstrated by decreased firing rate after iontophoretically-applied dopamine or systemically-administered apomorphine. These presynaptic effects were blocked by pretreatment with chronic lithium [*Gallager and Bunney,* 1979].

Data compatible with the above observations were reported by *Allikmets* et al. [1979] using an aggression model in rats. Rats withdrawn from chronic haloperidol exhibited a supersensitive aggressive response to apomorphine. Those animals pretreated with lithium did not develop supersensitivity.

Friedman et al. [1979] used reserpine-induced supersensitivity and employed apomorphine to induce stereotypic behavior in rats. They showed that lithium pretreatment enhanced rather than decreased reserpine-elicited sensitivity in the dopamine system. The authors suggested that the contrast in findings may be explained by the different mechanisms of action of haloperidol and reserpine.

Another model was utilized by *Seeger and Gardner* [1979], who demonstrated that chronic haloperidol treatment for 21 days was associated with an increase in intracranial self-stimulation rate in rats implanted with electrodes in the ventral tegmental area. This increase persisted for 3 weeks at full magnitude before declining to baseline at 4 weeks after the termination of haloperidol treatment. Chronic lithium treatment significantly attenuated both the magnitude and the time course of the neuroleptic-induced supersensitivity.

These results demonstrate the reversal of supersensitivity in a behavior mediated by the mesolimbic dopamine system and are compatible with findings of *Pert* et al. [1978] in the nigrostriatal dopamine system.

Beta-Receptors

Lithium treatment alone has been shown to decrease beta-receptor (^{3}H-DHA) binding [*Pert* et al., 1979; *Treiser and Kellar,* 1979; *Rosenblatt* et al., 1979]. In combination with other drugs lithium was demonstrated to block the development of 6-hydroxydopamine-induced beta-receptor supersensitivity [*Pert* et al., 1979] as well as reserpine-induced beta-receptor supersensitivity [*Treiser and Kellar,* 1979]. It is generally reported that many psychoactive compounds produce subsensitive beta-receptors [*Banerjee* et al., 1977; *U'Prichard and Enna,* 1979; *Rosenblatt* et al., 1979; *Sarai* et al., in press]. Lithium has failed to block the development of beta-receptor subsensitivity induced by chronic imipramine [*Rosenblatt* et al., 1979].

Alpha-Receptors

Studies of the effects of lithium alone on alterations in the sensitivity of alpha-receptors have produced less consistent findings. *Rosenblatt* et al. [1979] have reported that lithium alone increases alpha-receptor binding while *Treiser and Kellar* [1979] failed to show any effect of lithium alone on alpha-receptors. 6-Hydroxydopamine-induced alpha-receptor supersensitivity is prevented by lithium treatment [*Pert* et al., 1979]. *U'Prichard* [unpublished] has demonstrated that alpha-receptor binding is increased by chronic imipramine treatment. He also noted that while alpha-adrenergic receptor supersensitivity is routinely observed after chronic amitriptyline administration, chronic imipramine administration gives less reproducible alterations of alpha-receptor binding. This may be compatible with the findings of *Rosenblatt* et al. [1979], since they were unable to demonstrate an imipramine-induced supersensitivity of alpha-adrenergic receptors following chronic treatment with imipramine. However, using a different model, *Treiser and Kellar* [1979] reported that lithium did not significantly affect the reserpine-induced binding of ^{3}H-WE-4101 to alpha-receptors.

In a recent paper, *Maggi* et al. [1980] demonstrated an interaction of alpha- and beta-receptors. They incubated rat cerebral cortical slices with isoproterenol, a beta-adrenergic agonist, which produced an increase in alpha-adrenergic receptor binding in addition to a decrease in beta-adrenergic receptor binding. They suggest that the inverse modulatory relationship may be a mechanism for the homeostatic control of central noradrenergic activity.

Serotonin Receptors

Lithium alone has no effect on serotonin receptor sensitivity [*Gallager* et al., 1978]. The possible specificity of lithium for the catecholamine-supersensitive receptors may be indicated by the report that lithium was unable to block serotonin supersensitivity induced by imipramine (as evaluated electrophysiologically. The ability of chronic lithium administration to modify tricyclic antidepressant-induced supersensitivity development in cells receiving 5-hydroxytryptamine input was studied using microiontophoretic techniques [*Gallager and Bunney,* 1979]. Clorimipramine or imipramine was administered for a period of 14 days and resulted in a 5-fold increase in the sensitivity of hippocampal pyramidal cells to iontophoretically-applied 5-hydroxytryptamine. Lithium failed to block this increase in sensitivity of these cells. Utilizing another system, *Friedman* et al. [1979] studied oxy-5-methoxy N,N-dimethyltryptamine-induced head twitches in mice, which were increased following reserpine administration. Chronic lithium also failed to block this reserpine-elicited response and, in fact, apparently potentiated the response. It is of interest that lithium is unable to block the imipramine or amitriptyline-induced manias observed in manic-depressive patients [*Bunney* et al., 1972].

AMPT as a Model of Manic-Depressive Mood Change

If one were to develop a possible drug model of depressive and hypomanic mood changes, one might list the following criteria: the drug would produce depression associated with decreased catecholamines; the point of discontinuation of the drug would be associated with hypomania and insomnia, with increased functional catecholamines accompanied by increased receptor sensitivity; and the hypomania would be treatable with lithium. *Sitaram* et al. [unpublished] have reported the utilization of AMPT, a relatively specific tyrosine hydroxylase inhibitor that decreases the synthesis of NE and dopamine, to study mood changes in normal individuals. In 6 out of 6 animal studies, receptor supersensitivity was documented following discontinuation of AMPT [*Geyer and Segal,* 1973; *Beuthin* et al., 1972; *Redmond* et al., 1971; *Tarsy and Baldessarini,* 1974; *Dominic and Moore,* 1969; *Perez-Cruet* et al., 1971]. When normal controls (who gave informed consent) received AMPT they developed brief but ratable depressive symptoms associated with evidence of decreased catecholamine synthesis. Following discontinuation of AMPT there was observable hypomania accompanied by insomnia. Furthermore, *Sitaram* [unpublished] has also recently shown that this hypomania in normal individuals can be treated with lithium carbonate. This drug model and the observations associated with it are compatible with the hypothesis developed by *Post and Bunney,* as discussed earlier in this paper. In the future, there will continue to be a search for ways of studying receptor function in man. The goal is to find a dose range for specific agonists and antagonists which alter specific whole-body behaviors in a manner useful for developing dose-respone curves. Post-mortem studies continue to be of considerable interest.

References

Allikmets, L.H.; Stanley, M.; Gershon, S.: The effect of lithium on chronic haloperidol enhanced apomorphine aggression in rats. Life Sci. *25:* 165–170 (1979).

Ary, M.; Lomax, P.; Cox, B.: Apomorphine hypothermia in the rat: central sites and mechanisms. Neuropharmacology *16:* 731–735 (1977).

Banerjee, S.P.; Kung, L.S.; Riggi, S.J.; Chanda, S.K.: Development of beta-adrenergic receptor subsensitivity by antidepressants. Nature, Lond. *268:* 455–456 (1977).

Beuthin, F.C.; Miya, T.S.; Blake, D.E.; Bousquet, W.F.: Enhanced sensitivity to noradrenergic agonists and tolerance development to alpha-methyltyrosine in the rat. J. Pharmac. exp. Ther. *181:* 446–456 (1972).

Briley, M.S.; Langer, S.Z.; Raisman, R.; Sechter, D.; Zarifian, E.: Tritiated imipramine binding sites are decreased in platelets of untreated depressed patients. Science *209:* 303–305 (1980).

Briley, M.S.; Raisman, R.; Langer, S.Z.: Human platelets possess high-affinity binding sites for ^{3}H-imipramine. Eur. J. Pharmacol. *58:* 347–348 (1979).

Brodie, H.K.H.; Murphy, D.L.; Goodwin, F.K.; Bunney, W.E., Jr.: Catecholamines and mania: the effect of alpha-methyl-para-tyrosine on manic behavior and catecholamine metabolism. Clin. Pharmacol. Ther. *12:* 218–224 (1971).

Bunney, W.E., Jr.: Neuronal receptor function in psychiatry: strategy and theory; in Usdin, Bunney, Davis, Neuroreceptors: basic and clinical aspects, pp. 241–256 (Wiley, New York 1981.

Bunney, W.E., Jr.; Goodwin, F.K.; Murphy, D.L.; House, K.M.; Gordon, E.K.: The 'switch process' in manic-depressive illness. II. Relationship to catecholamines, REM sleep and drugs. Archs gen. Psychiat. *27:* 304–309 (1972).

Bunney, W.E., Jr.; Kammen, D.P. van; Post, R.M.; Garland, B.L.: A possible role for dopamine in schizophrenia and manic-depressive illness. (A review of evidence); in Usdin, Kopin, Barchas, Catecholamines: basic and clinical frontiers, pp. 1807–1819 (Pergamon Press, New York 1979).

Bunney, W.E., Jr.; Kopanda, R.T.; Murphy, D.L.: Sleep and behavioral change possibly reflecting central receptor hypersensitivity following catecholamine synthesis inhibition in man. Acta psychiat. scand. *56:* 189–203 (1977a).

Bunney, W.E., Jr.; Post, R.M.; Andersen, A.E.; Kopanda, R.T.: A neuronal receptor sensitivity mechanism in affective illness: a review of evidence. Commun. Psychopharmacol. *1:* 393–405 (1977b).

Burt, S.E.; Creese, I.; Snyder, S.H.: Antischizophrenic drugs: chronic treatment elevates dopamine receptor binding in brain. Science *196:* 326–328 (1977).

Cohen, R.M.; Campbell, I.C.; Cohen, M.R.; Torda, T.; Pickar, D.; Siever, L.J.; Murphy, D.L.: Presynaptic noradrenergic regulation during depression and antidepressant drug treatment. Psychiat. Res. *3:* 93–105 (1980).

Cox, B.; Lee, T.F.: Do central dopamine receptors have a physiological role in thermoregulation. Br. J. Pharmacol. *61:* 83–86 (1977).

Cox, B.; Lee, T.F.: Effect of central injections of dopamine on core temperature and thermoregulatory behaviour in unrestrained rats. Neuropharmacology *18:* 537–540 (1979).

Creese, I.; Burt, D.R.; Snyder, S.H.: Dopamine receptor binding enhancement accompanies lesion-induced behavioral supersensitivity. Science *197:* 596–598 (1977).

Cutler, N.R.; Post, R.M.; Bunney, W.E., Jr.: Apomorphine: brain dopaminergic function in man. Proc., 132nd Annu. Meet. Psychiat. Ass., Chicago, new Research Abstracts No. 31.

Dominic, J.A.; Moore, K.E.: Supersensitivity to the central stimulant actions of adrenergic drugs following discontinuation of a chronic diet of alpha-methyltyrosine. Psychopharmacologia *15:* 96–101 (1969).

Donoso, A.O.; Bishop, W.; Fawcett, C.P.; Krulich, L.; McCann, S.M.: Effect of drugs that modify brain monoamine concentrations on plasma gonadotropin and prolactin levels in the rat. Endocrinology *89:* 774–784 (1971).

Extein, I.; Tallman, J.; Smith, C.; Goodwin, F.: Changes in lymphocyte beta-adrenergic receptors in depression and mania. Psychiat. Res. *1:* 191–197 (1979).

Friedman, E.; Daliob, A.; Levine, G.: The effect of long-term lithium treatment on reserpine-induced supersensitivity in dopaminergic and serotonergic transmission. Life Sci. *25:* 1263–1266 (1979).

Gallager, D.W.; Bunney, W.E.; Jr.: Failure of chronic lithium treatment to block tricyclic antidepressant-induced 5HT supersensitivity. Archs Pharmacol. *307:* 129–133 (1979).

Gallager, D.W.; Pert, A.; Bunney, W.E., Jr.: Haloperidol-induced presynaptic dopamine supersensitivity is blocked by chronic lithium. Nature, Lond. *273:* 309–312 (1978).

Gerner, R.H.; Post, R.M.; Bunney, W.E., Jr.: A dopaminergic mechanism in mania. Am. J. Psychiat. *133:* 1177–1180 (1976).

Geyer, M.A.; Segal, D.S.: Differential effects of reserpine and alpha-methyl-para-tyrosine on norepinephrine and dopamine-induced behavioral activity. Psychopharmacologia *29:* 131–140 (1973).

Ishikawa, S.; Oono, S.; Hikita, H.: Drugs affecting the iris muscle; in Dikstein, Drugs and ocular tissues, chapter 5. 2nd Meet. Int. Soc. Eye Research, Jerusalem, pp. 288–382 (Karger, Basel 1976).

Jimerson, D.C.; Lynch, H.J.; Post, R.M.; Wurtman, R.J.; Bunney, W.E., Jr.: Urinary melatonin rhythms during sleep deprivation in depressed patients and normals. Life Sci. *20:* 1501–1508 (1977).

Kamberi, L.A.; Mical, R.; Porter, J.C.: Effect of anterior pituitary perfusion and intraventricular injection of catecholamines on prolactin release. Endocrinology *88:* 1012–1020 (1971).

Ketchum, J.S.; Sidell, F.R.; Crowell, E.; Aghajanian, G.K.; Hayes, A.H.: Atropine, scopolamine and citran; comparative pharmacology and antagonists in man. Psychopharmacology *28:* 121 (1973).

Lewy, A.J.; Wehr, T.A.; Gold, P.W.; Goodwin, F.K.: Plasma melatonin in manic-depressive illness; in Usdin, Kopin, Barchas, Catecholamines: basic and clinical frontiers, pp. 1173–1175 (Pergamon Press, New York 1979).

Lynch, H.J.; Jimerson, D.C.; Ozaki, Y.; Post, R.M.; Bunney, W.E., Jr.; Wurtman, R.J.: Entrainment of rhythmic melatonin secretion in man to a 12-hour phase shift in the light/dark cycle. Life Sci. *23:* 1557–1564 (1978).

Lynch, H.J.; Wurtman, R.J.; Moskowitz, M.A.; Archer, M.C.; Ho, M.H.: Daily rhythm in human urinary melatonin. Science *187:* 169–171 (1975).

Maggi, A.; U'Prichard, D.C.; Enna, S.J.: Beta-adrenergic regulation of alpha-adrenergic receptors in the central nervous system. Science *207:* 645–647 (1980).

Mapstone, R.: Safe mydriasis. Br. J. Ophthal. *54:* 690–692 (1976).

Meltzer, H.Y.; Goode, D.J.; Fang, V.S.: The effect of psychotropic drugs on endocrine function. I. Neuroleptics, precursors, agonists; in Lipton, DiMascio, Killam, Psychopharmacology: a generation of progress, pp. 509–529 (Raven Press, New York 1978).

Meltzer, H.Y.; Sachar, E.J.; Frantz, A.G.: Serum prolactin levels in acutely psychotic patients: an indirect measurement of central dopaminergic activity; in Usdin, Neuropsychopharmacology of monoamines and their regulatory enzymes, pp. 299–316 (Raven Press, New York 1974).

Muller, P.; Seeman, P.: Brain neurotransmitter receptors after long-term haloperidol. Life Sci. *21:* 1751–1758 (1978).

Murphy, D.L.; Brodie, H.K.H.; Goodwin, F.K.; Bunney, W.E., Jr.: Regular induction of hypomania by *L*-dopa in bipolar manic-depressive patients. Nature, Lond. *229:* 135–136 (1971).

Pandey, G.M.; Dysken, M.W.; Garver, D.L.; Davis, J.M.: Beta-adrenergic receptor function in affective illness. Am. J. Psychiat. *136:* 675–678 (1979).

Pandey, G.N.; Garver, D.L.; Tamminga, C.; Ericksen, S.; Ali, S.I.; Davis, J.M.: Postsynaptic supersensitivity in schizophrenia. Am. J. Psychiat. *134:* 518–522 (1977).

Paul, S.M.; Rehavi, M.; Skolnick, P.; Goodwin, F.K.: Demonstration of specific 'high affinity' binding sites for ^{3}H-imipramine on human platelets. Life Sci. *26:* 953–959 (1980).

Perez-Cruet, J.; Murphy, D.L.; Bunney, W.E., Jr.: Changes in synthesis rate of serotonin, dopamine and norepinephrine after withdrawal of chronic treatment with alpha-methyl-tyrosine (aMT) in rats. Clin. Res. *19:* 735 (1971).

Pert, A.; Rosenblatt, J.E.; Sivit, C., Pert, C.B.; Bunney, W.E., Jr.: Long-term treatment with lithium prevents the development of dopamine receptor supersensitivity. Science *201:* 171–173 (1978).

Pert, C.B.; Pert, A.; Rosenblatt, J.E.; Tallman, J.F.; Bunney, W.E., Jr.: Catecholamine receptor stabilization: a possible mode of lithium's anti-manic action; in Usdin, Kopin, Barchas, Catecholamines: basic and clinical frontiers, pp. 583–585 (Pergamon Press, New York 1979).

Post, R.M.; Gerner, R.H.; Carman, J.S.; Bunney, W.E., Jr.: Effects of low doses of a dopamine receptor stimulator in mania. Lancet *i:* 203–204 (1976).

Post, R.M.; Gerner, R.H.; Carman, J.S.; Gillin, J.C.; Jimerson, D.C.; Goodwin, F.K.; Bunney, W.E., Jr.: Effects of a dopamine agonist piribedil in depressed patients. Archs gen. Psychiat. *35:* 609–615 (1978a).

Post, R.M.; Jimerson, D.C.; Bunney, W.E., Jr.; Goodwin, F.K.: Dopamine and mania: behavioral and biochemical effects of the dopamine receptor blocker pimozide. Psychopharmacology *67:* 297–305 (1980).

Post, R.M.; Lake, C.R.; Jimerson, D.C.; Bunney, W.E., Jr.; Wood, J.H.; Ziegler, M.G.; Goodwin, F.K.: Cerebrospinal fluid norepinephrine in affective illness. Am. J. Psychiat. *135:* 907–912 (1978b).

Prange, A.J.; Loosen, P.T.; Wilson, I.C.; Meltzer, H.Y.; Fang, V.S.: Behavioral and endocrine responses of schizophrenic patients to TRH (protirelin). Archs gen. Psychiat. *36:* 1086–1093 (1979).

Quock, R.M.; Gale, C.C.: Hypothermia-mediating dopamine receptors in the preoptic anterior hypothalamus of the cat. Archs Pharmacol. *285:* 297–300 (1974).

Raisman, R.; Briley, M.; Langer, S.Z.: Specific tricyclic antidepressant binding sites in rat brain. Nature, Lond. *281:* 148–150 (1979).

Redmond, D.E., Jr.; Maas, J.W.; Kling, A.; Dekirmenjian, H.: Changes in primate social behavior after treatment with alpha-methyl-para-tyrosine. Psychosom. Med. *33:* 97–113 (1971).

Rehavi, M.; Paul, S.M.; Skolnick, P.; Goodwin, F.K.: Demonstration of specific high affinity binding sites for [^{3}H] imipramine in human brain. Life Sci. *26:* 2273–2279 (1980).

Rosenblatt, J.E.; Pert, C.B.; Tallman, J.F.; Pert, A.; Bunney, W.E., Jr.: The effect of imipramine and lithium on alpha and beta-receptor binding in rat brain. Brain Res. *160:* 186–191 (1979).

Sachar, E.: Neuroendocrine disturbances in the affective disorders and schizophrenia. Meet. Am. Coll. Neuropsychopharm., Palm Springs 1973).

Sack, R.L.; Goodwin, F.K.: Inhibition of dopamine-beta-hydroxylase in manic patients. Archs gen. Psychiat. *31:* 649–658 (1974).

Sagales, T.; Grill, S.; Domino, E.F.: Differential effects of scopolamine and chlorpromazine on REM and NREM sleep in normal male subjects. Clin. Pharmacol. Ther. *10:* 522 (1968).

Sagales, T.; Grill, S.; Domino, E.F.: Effects of repeated doses of scopolamine on the electroencephalographic stages of sleep in normal volunteers. Clin. Pharmacol. Ther. *18:* 727 (1975).

Sarai, K.; Frazer, A.; Brunswick, D.; Mendels, J.: Desmethylimipramine-induced decrease in beta-adrenergic receptor binding in rat cerebral cortex. Biochem. Pharmac. (in press).

Schelkunov, E.L.: Action of neuroleptics and antidepressants on the hypothermia produced by dopamino-mimetics apomorphine and piribedil in mice. J. neural Transm. *47:* 307–312 (1980).

Seeger, T.F.; Gardner, E.L.: Reversal of dopaminergic behavioral supersensitivity: effects of chronic *L*-dopa and lithium. Abstr. Soc. Neurosci. No. 2259, pp. 662 (1979).

Sitaram, N.; Gillin, J.C.: Development and use of pharmacological probes of the CNS in man: evidence of cholinergic abnormality in primary affective illness. Biol. Psychiat. (in press).

Sitaram, N.; Moore, A.M.; Gillin, J.C.: Scopolamine-induced muscarinic supersensitivity in normal man: changes in sleep. Psychiat. Res. *1:* 9–16 (1979).

Smith, S.A.; Smith, S.E.: Factors determining the potency of cholinomimetic miotic drug and their effect upon the light reflex in man. Br. J. Pharmacol. *6:* 149–153 (1978).

Smith, R.C.; Tamminga, C.; Haraszti, J.; Pandey, G.M.; Davis, J.M.: Effects of dopamine agonists in tardive dyskinesia. Am. J. Psychiat. *134:* 763–768 (1977).

Tarsy, D.; Baldessarini, R.J.: Behavioral supersensitivity to apomorphine following chronic treatment with drugs which interfere with the synaptic function of catecholamines. Neuropharmacology *13:* 927–940 (1974).

Treiser, S., and Kellar, K.J.: Lithium effects on adrenergic receptor supersensitivity in rat brain. Eur. J. Pharmacol. *58:* 85–86 (1979).

U'Prichard, D.C.; Enna, S.J.: In vitro modulation of CNS beta-receptor number by antidepressants and beta-agonists. Eur. J. Pharmacol. *59:* 297–302 (1979).

W.E. Bunney, Jr., MD, Biological Psychiatry Branch, National Institute of Mental Health, Bethesda, MD 20205 (USA)

Adv. biol. Psychiat., vol. 7, pp. 85–89 (Karger, Basel 1981)

Platelet Alpha-Adrenoreceptor Sensitivity in Depressive Illness

Keith Wood, Alec Coppen

Medical Research Council Neuropsychiatry Laboratory, West Park Hospital, Epsom, Surrey, England

Introduction

The therapeutic efficacy of certain tricyclic compounds to inhibit the neuronal reuptake of released noradrenaline (NA) and 5-hydroxytryptamine (5-HT) has been one of the major pieces of experimental evidence to sustain the monoamine hypothesis of depression [*Schildkraut,* 1965; *Coppen,* 1967]. The relative or absolute deficiency of these amines within the central nervous system of depressive patients may be accounted for by defects in the synthesis, transport, compartmentation, release, reuptake and post-synaptic receptor binding. However, recent investigations have failed to detect any significant relationships between tricyclic antidepressant drug induced inhibition of NA [*Ghose and Coppen,* 1977], and 5-HT reuptake [*Coppen* et al., 1979], and clinical improvement in depressed patients. Peripheral α-adrenoceptors have been found to be supersensitive during a depressive illness and a significant negative correlation between clinical improvement and a reduction in the sensitivity of these receptors has been established [*Coppen and Ghose,* 1978]. The demonstration of α-adrenoceptors on the surface of blood platelets [*Newman* et al., 1978], now allows direct studies to be made on these easily obtainable receptor sites during a depressive illness.

Methods

Patients and Controls

Female patients diagnosed to be suffering from primary depressive illness were studied. These patients were assessed for severity of depression by the first 16 items of the Hamilton Rating Scale [*Hamilton,* 1967]. Only patients who scored 16 or more on this scale after a 7-to-10-day drug-free assessment phase were included in this study. At this time the patients were assessed using the Newcastle diagnosis scale [*Carney* et al., 1965]. On this diagnosis scale high scores indicate an endogenous type of depression and lower scores neurotic depression. Women who volunteered to act as control subjects had no known history of psychiatric or physical illness nor had taken any medication 7–10 days prior to testing.

α-Adrenergic Receptor Binding Assay

Blood was collected from controls and patients after overnight fasting into 3.2% (w/v) sodium citrate and α-adrenergic receptors on intact platelets were measured with ^{3}H-dihydroergocryptine (^{3}H-DHE; The Radiochemical Centre, Ltd., Amersham, UK), according to previously published methods [*Wood and Coppen,* 1980]. Specific binding of ^{3}H-DHE was calculated as the difference between that bound in the presence and absence of 10 μmol/l phentolamine. Results from the experiments were calculated by a *Scatchard* [1949] plot by linear regression analysis according to the method of least squares and the patient and control group were compared for receptor affinity and capacity with the Student's unpaired t-test.

Results

The equilibrium dissociation constant (K_d; nmol/l) and the capacity of the site (B_{max}; receptors/platelet) were obtained by Scatchard plot analysis and are shown in table I. There was no significant difference in the values of K_d in any of the groups studied. There were significantly fewer ^{3}H-DHE sites on intact platelets of depressed patients when compared to the post-menopausal controls. There was a trend for patients with a Newcastle score of ≥ 6 to have fewer ^{3}H-DHE sites than patients with a Newcastle score of ≤ 5 but the difference was not significant.

Discussion

In contrast to the results of the pressor response tests [*Coppen and Ghose,* 1978] ^{3}H-DHE binding assays have demonstrated a reduced number of α-receptors and possible subsensitivity of the platelet system in depressed patients even though both tests are measures of α-adrenoceptor sensitivity. This discrepency may be accounted for by two factors: (a) that the pressor response tests are measures of vascular reactivity and not of α-adrenergic sensitivity and this can only be established by determining whether hypersensitivity to angiotensin in terms of a blood pressure rise occurs, and (b) the presence of two different varieties of these α-receptors in the periphery. Recent pharmacological studies have shown that some compounds exhibit a preferential specificity for one of the two types of receptors. These characteristics represent the basis of a classification of α_1- and α_2-adrenoceptors. In the periphery, α_1-adrenoceptors correspond to the postsynaptic receptors and α_2-adrenoceptors to the presynaptic ones. It has been suggested that ^{3}H-DHE is a non-selective antagonist in the platelet system and

Table I. ^{3}H-dihydroergocryptine binding characteristics (K_d and B_{max}) in post-menopausal controls and female depressed patient (results presented as means ± SEM)

Group	n	Age, years	K_d, (nm)	B_{max}, (receptors/platelet)
Post-menopausal controls	9	56.1 ± 2.5	2.6 ± 0.2	296 ± 13
All depressed patients	14	61.5 ± 2.3	2.8 ± 0.1	203 ± 14**
Newcastle ⩽5	4	56.3 ± 4.1	2.7 ± 0.3	238 ± 20*
Newcastle ⩾6	10	63.6 ± 2.7	2.8 ± 0.2	190 ± 17**

Significantly lower than controls: * $p < 0.05$; ** $p < 0.001$.

will therefore label both α_1 and α_2 sites which are present on the platelet membrane [*Scrutton and Grant,* 1979]. Recent evidence suggests that the majority of platelet α-receptors are of the α_2 type although there appears to be a great deal of interindividual variation in the proportion of these two types of receptors [*Scrutton,* personal commun.]. The discrepancy between the results of the pressor response tests and ^{3}H-DHE binding assays may possibly be explained by the relative efficiency in distinguishing between α_1- and α_2-receptors, respectively.

There are several explanations as to the reason for a reduced number of ^{3}H-DHE binding sites in the platelet membranes of depressed patients. It has been known for some time that a wide variety of hormones and drugs have the ability to regulate the concentration of their own receptors in tissues. In general there is an inverse relationship between the concentration of the biologically active molecule and the number of its receptors. Thus, high concentrations of a biologically active molecule seem to decrease or 'down-regulate' receptor number and in contrast receptor density may increase when the concentrations of the active molecule are lowered. Adrenergic catecholamines, oestrogen, denervation, hyperthyroidism and thrombocystosis have 'up' and 'down' regulating effects upon the α-adrenergic receptor. If the plasma concentrations of endogenous α-agonists are higher in depressed patients then the observed decreased binding of ^{3}H-DHE to platelets might be expected to occur in depressed patients. Increased plasma catecholamine levels have been found in conditions of stress and anxiety [*Frankenhaeuser,* 1975] and *Wyatt* et al. [1971] have reported that total resting plasma catecholamine concentrations of depressed patients are higher than in controls. It is interesting to note that there is a trend for patients with a Newcastle score of ⩾6 to have a lower number of receptors than the patients with a Newcastle score of ⩽5 and this may be related to the increased urinary excretion of cortisol in patients with a Newcastle score of ⩾6 compared to those with a score of ⩽5 [*Coppen* et al., unpublished observation]. Therefore, until further experimentation has been completed to determine the cause of the down regulation of platelet α-receptors it is a little premature to assume that parallel changes occur in the brains of depressive patients. Further animal experiments should explore the validity of extrapolating from platelets to the brain by investigating the relationship between ^{3}H-DHE binding in the brain and platelets of the same experimental animal.

Plasma platelet populations are determined to some extent by the regulation of the exchangeable splenic platelet pool which is mediated by β-receptors [*Fredén* et al., 1979] which have been reported to be down-regulated in the periphery of depressed patients [*Extein* et al., 1980], and abnormalities of this β-receptor in depressive illness may produce an alteration in the type of platelet released into the circulation.

Any process affecting protein synthesis in the megakaryocyte, the nucleated platelet stem cell, must similarly affect the number of receptors present on the surface of blood platelets. Investigations into platelet synthesis and causes of down-regulation of the platelet α-adrenoceptor must be elucidated before we can interpret the results presented in this investigation as a change occurring in the central nervous system of depressed patients.

Summary

α-Adrenoreceptor sensitivity of blood platelets of depressed, drug-free patients and controls were determined using a radio-ligand binding assay with ^{3}H-dihydroergocryptine. Depressed patients were found to have a significantly reduced number of α-adrenoreceptor sites (B_{max}) when compared to the control group although no such difference was detected in the dissociation constant (K_d). The results are discussed with reference to the regulation of α-adrenoreceptors and possible parallel changes occurring in the central nervous system of depressed patients.

References

Carney, M.W.P.; Roth, M.; Garside, R.F.: The diagnosis of depressive syndromes and the prediction of ECT response. Br. J. Psychiat. *111:* 659–674 (1965).

Coppen, A.: The biochemistry of affective disorders. Br. J. Psychiat. *113:* 1237–1264 (1967).

Coppen, A.; Ghose, K.: Peripheral α-adrenoreceptor and central dopamine receptor activity in depressive patients. Psychopharmacology *59:* 171–177 (1978).

Coppen, A.; Rao, V.A.R.; Swade, C.; Wood, K.: Inhibition of 5-hydroxytryptamine reuptake by amitriptyline and zimelidine and its relationship to their therapeutic action. Psychopharmacology *63:* 125–129 (1979).

Extein, I.; Tallman, J.; Smith, C.C.; Goodwin, F.K.: Changes in lymphocyte β-adrenergic receptors in depression and mania. Psychiat. Res. (in press, 1980).

Frankenhaeuser, M.: Experimental approaches to the study of catecholamines and emotions; in Levi, Emotions: their parameters and measurement, pp. 209–234 (Raven Press, New York 1975).

Fredén, K.; Vilén, L.; Lundborg, P.; Olsson, L.-B.; Kutti, J.: The peripheral platelet count and the isoprenaline-induced splenic platelet pooling in response to β-adrenoceptor blockade. Scand. J. Haematol. *23:* 245–250 (1979).

Ghose, K.; Coppen, A.: Noradrenaline, depressive illness, and the action of amitriptyline. Psychopharmacology *54:* 57–60 (1977).

Hamilton, M.: Development of a rating scale for primary depressive illness. Br. J. soc. clin. Psychol. *6:* 278–296 (1967).

Newman, K.D.; Williams, L.T.; Bishopric, N.H.; Lefkowitz, R.J.: Identification of α-adrenergic receptors in human platelets by ^{3}H-dihydroergocryptine binding. J. clin. Invest. *61:* 395–402 (1978).

Scatchard, G.: The attractions of proteins for small molecules and ions. Ann. N.Y. Acad. Sci. *51:* 660–672 (1949).

Schildkraut, J.J.: The catecholamine hypothesis of affective disorders: a review of the supporting evidence. Am. J. Psychiat. *122:* 509–522 (1965).

Scrutton, M.C.; Grant, J.A.: Dihydroergocryptine is a non-selective antagonist for human platelet α-adrenoreceptors. Nature, Lond. *280:* 700 (1979).

Wood, K.; Coppen, A.: Hormonal influences on α-adrenoreceptors: preliminary results; in Pasetto, Paoletti, Ambrus, The menopause and postmenopause, (MTP Press Ltd., Lancaster 1980).

Wyatt, R.J.; Portnoy, B.; Kupfer, D.J., Snyder, F.; Engelman, K.: Resting plasma catecholamine concentrations in patients with depression and anxiety. Archs gen. Psychiat. *24:* 65–70 (1971).

K. Wood, PhD, Medical Research Council Neuropsychiatry Laboratory,
West Park Hospital, Epsom, KT19 8PB (England)

Adv. biol. Psychiat., vol. 7, pp. 90–99 (Karger, Basel 1981)

New Perspectives on the Action of Antidepressant drugs: Regulation of Central Adrenergic Receptor Funktion[1]

Fridolin Sulser

Vanderbilt University School of Medicine and Tennessee Neuropsychiatric Institute, Nashville, Tenn., USA

Introduction

The purpose of this paper is to focus attention on alterations in the sensitivity of central adrenergic receptor systems as an important regulatory control of a specific biologic response elicited by the neurohormone norepinephrine (NE). Studies from many laboratories have clearly shown that receptors are not constant elements in the cell membrane and that the density of receptors and the sensitivity of receptor systems change with the change of the cell cycle, during cellular development and differentiation, as a function of time of the day (circadian rhythm) and under the influence of various hormones and drugs.

The NE receptor coupled adenylate cyclase system in the limbic forebrain, cortex and in other areas of the brain with noradrenergic projections is particularly suited to highlight some general aspects of adrenergic receptor regulation and receptor plasticity. This system shows all the characteristics of a functional receptor system including recognition function and action function. Moreover, psychotropic drugs which can precipitate or alleviate depression regulate this particular receptor system in opposite directions. Table I lists some of the characteristics of this system including stereospecificity for agonists and antagonists and the development of supersensitivity following chemical sympathectomy or surgical denervation [*Kalisker* et al., 1973; *Blumberg* et al., 1976; *Robinson* et al., 1978; *Mobley and Sulser,* 1979; *Mishra* et al., 1981]. Since in binding studies, it is often a problem to determine whether the site detected by the ligand represents a 'true' pharmacological receptor – even if the criteria of stereospecificity, reversibility and saturation binding are experimentally met – a receptor-linked biological response (action function of the system) is essential to distin-

[1] Research from this laboratory has been supported by USPHS grants MH-11468, MH-29228 and by the Tennessee Department of Mental Health and Mental Retardation.

Table I. NE receptor-coupled adenylate cyclase system in brain as a functional receptor system.

Transmitter specificity for NE and epinephrine (dopamine and serotonin are inactive)
Stereospecificity – β-OH group in *R* configuration
β-Adrenergic receptors represent subpopulation of NE receptors coupled to adenylate cyclase
α-Adrenergic receptors are not coupled to adenylate cyclase and consequently α-adrenergic agonists (e.g. phenylephrine, methoxamine) do not stimulate the system
Stereospecific blockade by (+)-butaclamol
Sensitivity of the system is up- or down-regulated by the receptor-specific agonist NE (homospecific regulation) and by other mechanisms (heterospecific regulation)

guish receptor-related binding from binding of ligands to cell surface components which may be termed 'acceptors'.

To better understand the multiple mechanisms which may be involved in sensitivity changes of the NE receptor coupled adenylate cyclase system, it is important to realize that the hormone-sensitive enzyme system generally consists of at least three physically separate proteins within the phospholipid bilayer of the membrane. Located at the outer surface of the membrane is the receptor containing a specific binding site for a particular hormone or neurohormone (recognition site). Adenylate cyclase and a nucleotide regulatory protein are located at the inner face of the membrane. The nucleotide regulatory protein contains sites for binding GTP. Before the discovery of the role of GTP as an essential activator of adenylate cyclase [*Rodbell* et al., 1971], it was thought that hormone receptors interacted directly with adenylate cyclase to form the second messenger cyclic AMP. It is now evident that the action of hormones by binding to the receptors is mediated through a complex coupling mechanism which involves nucleotide regulatory proteins that bind GTP [*Rodbell,* 1980].

Effect of Antidepressant Treatments on the Sensitivity of the NE Receptor Coupled Adenylate Cyclase System in Brain

The classical catecholamine and/or indoleamine hypothesis of affective disorders [*Schildkraut,* 1965; *Coppen,* 1972] have stimulated considerable research on affective disorders and had thus considerable heuristic value. However, the pharmacological basis of these hypotheses has been chiefly derived from studies on acute pharmacological effects elicited by a number of psychotropic drugs. Closer inspection of the pharmacological evidence for these hypotheses raises a number of questions as some preclinical and clinical findings are difficult to reconcile with these original hypotheses: (1) There exists a marked discrepancy in the time course between biochemical and pharmacological effects elicited by e.g. tricyclic antidepressants within minutes and their clinical therapeutic action which requires treatment for several weeks. (2) Marked variations in clinical responses occur despite a uniform pharmacologic inhibition of reuptake of either catechol- or indoleamines. (3) Drugs such as iprindole and mianserin which do not block the neuronal reuptake of NE or serotonin (5HT) are clinically effective (mianserin actually blocks central 5HT receptors). (4) Cocaine, a potent blocker of NE reuptake, is not an effective antidepressant. (5) *L*-Dopa, a precursor of NE, fails to rapidly reverse depressive symptomatology and amphetamine which is known to share pharmacologic properties with both MAO inhibitors and tricyclic antidepressants is a rather poor antidepressant drug in man. (6) Many of the treatments for depression appear to alter amine metabolism in the same direction which has been demonstrated or hypothesized to be a characteristic of the illness itself.

Such discrepancies prompted us to study the effect of antidepressants following their administration on a clinically more relevant time basis on the NE receptor-coupled adenylate cyclase system in brain. We found that chronic but not acute administration of various prototypes of antidepressant treatments caused a subsensitivity of this postsynaptic noradrenergic cyclic AMP generating system [*Vetulani and Sulser,* 1975; *Vetulani* et al., 1976a, b; *Mishra* et al., 1979; 1980]. Importantly, this delayed subsensitivity is not related to the concentration of the drugs in brain tissue but solely depends on time and it is also induced by clinically effective antidepressants which do not block the reuptake of NE [*Vetulani and Sulser,* 1975; *Vetulani* et al., 1976b, *Mishra* et al., 1980]. Since ECT is one of the most effective treatments of severe depression and its onset of action is generally considered to be more rapid than that of pharmacotherapy, it is of considerable interest that ECT elicited a very rapid subsensitivity of the cyclic AMP generating system to NE and to the β-adrenergic agonist isoproterenol [*Gillespie* et al., 1979]. Table II represents the growing list of antidepressant treatments which have so far been shown to elicit subsensitivity of this noradrenergic receptor system.

A change in the availability of 5HT per se does not alter the sensitivity of this noradrenergic receptor system as fluoxetine, the hitherto most specific 5HT uptake inhibitor (increased availability of 5HT) and raphe lesions (decrease in the availability of 5HT) were without effect [*Mishra* et al., 1981]. Since a persistent increase in the availability of NE appears to be essential for the development of subsensitivity [*Wolfe* et al., 1978; *Mishra* et al., 1979; *Manier* et al., 1980], we interpret the action on the noradrenergic receptor system by the more specific 5HT uptake inhibitors amitriptyline and chlorimipramine as being the consequence of their in vivo conversion to the corresponding secondary amines

Table II. Antidepressant treatments which elicit subsensitivity of the NE receptor-coupled adenylate cyclase system in brain

	Subsensitivity to NE and/or isoproterenol	Reduction in the density of β-adrenergic receptors
Antidepressant drugs which block uptake of 5HT and/or NE		
Chlorimipramine	+++	++
Imipramine	+++	++
Amitriptyline	+++	++
Antidepressant drugs which block predominantly uptake of NE		
Desipramine	+++	++
Nisoxetine	+++	0
Drugs which block selectively uptake of 5HT		
Fluoxetine	0	0
Zimelidine	+++	++
Antidepressant drugs which do not block uptake of NE and/or 5HT		
Iprindole	+++	++
Mianserin	+++	0
Antidepressant drugs which block MAO		
Pargyline	+++	++
Nialamide	+++	++
Tranylcypromine	+++	++
Electroconvulsive treatment	+++	++
Chlorpromazine	+++	0

The data have been obtained either in the author's laboratory or have been computed from the current literature [adapted from *Sulser,* 1979].

which are known to be more potent inhibitors of the reuptake of NE. *Schultz* [1976] reported that chlorpromazine is also down-regulating noradrenergic receptor sensitivity following chronic administration of the drug. Since chlorpromazine blocks the neuronal reuptake of NE and also blocks α_2-adrenergic receptors (both actions would tend to increase the availability of NE), it is not surprising that this drug also elicits subsensitivity of the NE receptor-coupled adenylate cyclase system. The effects of antidepressant drugs elicited on α_2-adrenergic receptors, either blockade (mianserin) or induction of subsensitivity (desipramine) are clearly of interest [*Fludder* et al., 1979; *Crews and Smith,* 1978] as these actions would increase the availability of NE and thus contribute to the NE-induced down-regulation of the adrenergic cyclic AMP generating system. It has recently been suggested that an acute blockade of histamine H_2-sensitive adenylate cyclase by antidepressant drugs might secondarily lead to a subsensitivity of the NE receptor-coupled adenylate cyclase system [*Kanof and Greengard,* 1978]. This acute effect elicited by antidepressant drugs on histamine-sensitive adenylate cyclase is however, unlike the development of subsensitivity to NE, not a common feature of all antidepressants.

The development of adrenergic subsensitivity following chronic treatment with antidepressants has been confirmed in a number of laboratories [*Schultz,* 1976; *Schmidt and Thornberry,* 1977; *Frazer* et al., 1978; *Wolfe* et al., 1978; *Korf* et al., 1979; *Moyer* et al., 1979] and electrophysiological correlates of noradrenergic subsensitivity have been established. Thus, a decreased responsiveness to iontophoretically applied NE has been reported in cerebellar Purkinje cells [*Siggins and Schultz,* 1979] and in neurons of the cingulate cortex [*Olpe and Schellenberg,* 1980]. In apparent contrast with these electrophysiological results, *Wang and Aghajanian* [1980] reported a marked increase of the responsiveness of amygdaloid neurons to both iontophoretically applied 5HT and NE following chronic treatment with tricyclic antidepressants and a potentiation of the inhibitory response of amygdaloid cells to stimulation of the ascending 5HT pathway. The reasons for these apparent discrepancies in electrophysiological results remain to be elucidated. In any event, in comparing the acute pharmacological effects of antidepressant treatments at presynaptic noradrenergic or serotonergic sites with those on noradrenergic receptor function following treatment on a clinically relevant time basis, it becomes apparent that the delayed therapeutic action of antidepressant treatments is better related in time to the delayed changes in adrenergic receptor function [*Sulser,* 1978].

The Role of the Density of Receptors in the Development of Functional Sensitivity Changes

Considering the complexity of the receptor-coupled adenylate cyclase system within the phospholipid bilayer of the membrane, some of the possible mechanisms responsible for subsensitivity of the NE receptor-coupled adenylate cyclase system include: (1) changes in the affinity of NE to the receptor; (2) reduction in the density of adrenergic receptors; (3) reversible uncoupling of the receptor and adenylate cyclase; (4) changes in the guanine nucleotide regulatory process; (5) changes in the catalytic activity of adenylate cyclase, and (6) changes in the activity of phosphodiesterase.

So far, studies on up- or down-regulation of the NE receptor-coupled adenylate cyclase system by psychotropic drugs have failed to reveal changes in the affinity of NE to its receptor or in the synthesis or degradation of the nucleotide by changing the activity of adenylate cyclase or phosphodiesterase. Considerable data exist, however, demonstrating that central noradrenergic subsensitivity elicited by antidepressant drugs and by ECT is linked to a decrease in the number of β-adrenergic receptors as assessed by binding studies with radioligands such as dihydroalprenolol or hydroxybenzylpindolol [*Banerjee* et al., 1977; *Wolfe* et al., 1978; *Sarai* et al., 1978; *Bergstrom and Kellar,* 1979a, b; *Gillespie* et al., 1979; *Pandey* etsal., 1979]. Scatchard analysis of such binding data indicates that the changes in the density of β-adrenergic receptors occur without changes in the K_d value. Moreover, these changes in the B_{max} value of β-adrenergic receptors appear to be quite specific for antidepressant treatments as no changes have been found following the administration of a variety of other CNS-active drugs [*Sellinger-Barnett* et al., 1980]. *Minneman* et al. [1979] have provided evidence that β_1-adrenergic receptors are down-regulated by antidepressants such as desipramine suggesting that the β_1-subpopulation of adrenergic receptors is primarily involved in central neuronal function.

Wirtz-Justice et al. [1980] have added a temporal dimension to adrenergic binding studies. They have demonstrated circadian rhythms of α- and β-adrenergic receptors in brain and that certain antidepressant drugs can alter – in addition to changes in the absolute number – the relative number of adrenergic receptors at a given time of the day. Binding studies with ligands labelling α-adrenergic and serotonergic receptor sites have also been conducted. However, results on the effect of chronic treatment with antidepressant drugs on the density of these receptors are rather conflicting at this time.

The question arises whether a reduction in the number of β-adrenergic receptors is a prerequisite for down-regulation of the NE receptor-coupled adenylate cyclase system and the induction of noradrenergic subsensitivity. Studies from our laboratories have now shown that subsensitivity induced by antidepressant treatments is not invariably linked to a reduction in the density of β-adrenergic receptors (table III). Thus, nisoxetine, mianserin and zimelidine caused upon chronic administration subsensitivity of the cyclic AMP generating system to NE while the density of β-adrenergic receptors remained essentially unchanged. These experiments indicate that an alteration of the number of β-adrenergic receptors represents only one mechanism by which the hormonal sensitivity to NE can be modified and also that changes in the sensitivity of the NE receptor-coupled adenylate cyclase system are a more sensitive indicator of functional changes in adrenergic receptor function than are alterations in the number of β-adrenergic receptors.

Clearly, drug-induced alterations of transmembrane regulation past the receptor will have to be investigated such as reversible uncoupling of adrenergic receptors and adenylate cyclase. The role of guanyl nucleotides such as GTP in the control of the transmission of the hormonal message from the receptor to the enzyme has been first recognized by *Rodbell* et al. [1971, 1974]. Also, changes

Table III. Noradrenergic subsensitivity without reduction in β-adrenergic receptor binding

	Response to 100 μM NE pmol cyclic AMP/mg protein ± SEM	Specific ^{3}H-DHA binding % of control binding
Control	61 ± 9	100 ± 11
DMI (15 mg/kg)	18 ± 2***	73 ± 7**
Mianserin (30 mg/kg)	30 ± 3**	101 ± 6
Zimelidine (30 mg/kg)	41 ± 3*	92 ± 6
Nisoxetine (20 mg/kg b.i.d.)	29 ± 2**	92 ± 9
Fluoxetine (10 mg/kg)	55 ± 8	104 ± 14

The drugs were administered i.p. to Sprague-Dawley rats for a period of 2 weeks. The cyclic AMP responses to NE and the specific β-adrenergic receptor binding were determined in slices or homogenates respectively of the cerebral cortex [from *Mishra* et al., 1979, 1980]. *$p < 0.02$; **$p < 0.01$; ***$p < 0.001$.

in the methylation of phospholipids in the membrane have been shown to affect membrane fluidity and coupling of β-adrenergic receptors with adenylate cyclase [*Hirata and Axelrod,* 1978; *Hirata* et al., 1979]. Such results emphasize the importance of lipids in the membrane for receptor-mediated transmission of the NE signal and of biological signals in general. Also, the role of agents other than NE in the transmembrane regulation of the biological signal remains to be pursued. For example, studies from our laboratories have shown that adrenal corticoids can regulate the sensitivity of the NE receptor-coupled adenylate cyclase system in brain without altering the density of β-adrenergic receptors [*Mobley and Sulser,* 1980]. Also, *Wagner* et al. [1979] reported that 17-α-ethynyl estradiol decreased β-adrenergic responses linked to a decrease in the number of β-adrenergic receptors in the cerebral cortex of ovariectomized rats. It thus appears that steroids either directly or indirectly can act as endocrine regulators of central noradrenergic sensitivity. The elucidation of their role in both pathogenesis and therapy of affective disorders remains a fascinating topic.

Conclusions

The research reviewed in this paper provides an experimental basis for a shift in emphasis on the mode of action of antidepressant treatments from acute effects elicited at presynaptic noradrenergic and serotonergic sites to a delayed postsynaptic modification of adrenergic receptor function.

As pointed out previously [*Sulser,* 1979], 'the findings that treatments which either precipitate or alleviate depressive illness alter the sensitivity of the

NE receptor-coupled adenylate cyclase system in cortical and subcortical structures of the brain in opposite directions (supersensitivity or subsensitivity) raise new questions on the psychobiology of depressive illness. One is tempted to advance the hypothesis that depression prone patients may suffer from an inability to regulate or more precisely to downregulate the central noradrenergic cyclic AMP-mediated information flow (inability to adapt optimally to neuronal input) and consequently that successful treatment with antidepressant drugs and ECT depends on the successful induction of subsensitivity of the NE receptor-coupled adenylate cyclase system'.

The demonstration of the plasticity of central NE receptor systems and of adaptive mechanisms of adrenergic transmembrane regulation should catalyze the elucidation of the complex neuronal processes that translate the initial recognition of the NE signal (recognition function of adrenergic receptors) into the final intraneuronal response (action function of adrenergic receptor systems).

References

Banerjee, L.P.; Kung, L.S.; Riggi, S.T.; Chanda, S.K.: Development of β-adrenergic receptor subsensitivity by antidepressants. Nature, Lond. *268:* 455–456 (1977).

Bergstrom, D.A.; Kellar, K.J.: Effect of electroconvulsive shock on monoaminergic receptor binding sites in rat brain, Nature, Lond. *278:* 464–466 (1979a).

Bergstrom, D.A.; Kellar, K.J.: Adrenergic and serotonergic receptor binding in rat brain after chronic desmethylimipramine treatment. J. Pharmac. exp. Ther. *209:* 256–261 (1979b).

Blumberg, J.B.; Vetulani, J.; Stawarz, R.J.; Sulser, F.: The noradrenergic cyclic AMP generating system in the limbic forebrain: pharmacological characterization and possible role in the mode of action of antipsychotics. Eur. J. Pharmacol. *37:* 357–366 (1976).

Coppen, A.: Indole amines and affective disorders. J. psychiat. Res. *9:* 163–171 (1972).

Crews, F.T.; Smith, C.B.: Presynaptic alpha-receptor subsensitivity after long-term antidepressant treatment, Science *202:* 322–324 (1978).

Fludder, G.M.; Leonard, B.E.: Chronic effects of mianserin on noradrenaline metabolism in the rat brain: evidence for a presynaptic α-adrenolytic action in vivo. Psychopharmacology *64:* 329–332 (1979).

Frazer, A.; Hess, M.E.; Mendels, J.; Gable, B.; Kunkel, E.; Bender, A.: Influence of acute and chronic treatment with desmethylimipramine on catecholamine effects in the rat. J. Pharmac. exp. Ther. *206:* 311–319 (1978).

Gillespie, D.D.; Manier, D.H.; Sulser, F.: Electroconvulsive treatment: rapid subsensitivity of the norepinephrine receptor coupled adenylate cyclase system in brain linked to downregulation of β-adrenergic receptors. Commun. Psychopharmacol. *3:* 191–195 (1979).

Hirata, F.; Axelrod, J.: Enzymatic methylation of phosphatidylethanolamine increases erythrocyte membrane fluidity. Nature, Lond. *275:* 219–220 (1978).

Hirata, F.; Strittmatter, W.J.; Axelrod, J.: β-Adrenergic receptor agonists increase phospholipid methylation, membrane fluidity and β-adrenergic receptor adenylate cyclase coupling. Proc. natn. Acad. Sci. USA *76:* 368–372 (1979).

Kalisker, A.; Rutledge, C.H.O.; Perkins, J.P.: Effect of nerve degeneration by 6-hydroxydopamine on catecholamine-stimulated adenosine 3′, 5′-monophosphate formation in rat cerebral cortex. Molec. Pharmacol. *9:* 619–629 (1973).

Kanof, P.O.; Greengard, P.: Brain histamine receptors as targets for antidepressant drugs. Nature, Lond. *272:* 329–333 (1978).

Korf, J.; Sebens, J.B.; Postema, F.: Cyclic AMP in the rat cerebral cortex after stimulation of the locus coeruleus. Decrease by antidepressent drugs. Eur. J. Pharmacol. *59:* 23–30 (1979).

Manier, D.H.; Gillespie, D.D.; Sulser, F.: Development of and recovery from subsensitivity of the noradrenergic cyclic AMP generating system in brain: effect of amphetamine following inhibition of its aromatic hydroxylation by iprindole. Arch. Pharmacol. *313:* 113–118 (1980).

Minneman, K.P.; Dibner, M.D.; Wolfe, B.B.; Molinoff, P.B.: β_1- and β_2-adrenergic receptors in rat cerebral cortex are independently regulated. Science *204:* 866–868 (1979).

Mishra, R.; Janowsky,. A.; Sulser, F.: Subsensitivity of the NE receptor coupled adenylate cyclase system in brain: effects of nisoxetine or fluoxetine. Eur. J. Pharmacol. *60:* 379–383 (1979).

Mishra, R.; Janowsky, A.; Sulser, F.: Action of mianserin and zimelidine on the norepinephrine receptor coupled adenylate cyclase system in brain: subsensitivity without reduction in β-adrenergic receptor binding. Neuropharmacology *19:* 983–987 (1980).

Mishra, R.; Leith, N.J.; Steranka, L.; Sulser, F.: Role of changes in the availability of serotonin (5HT) in the development of subsensitivity of the norepinephrine receptor coupled adenylate cyclase system in brain. Arch. Pharmacol. (in press, 1981).

Mobley, P.L.; Sulser, F.: Norepinephrine stimulated cyclic AMP accumulation in rat limbic forebrain slices: partial mediation by a subpopulation of receptors with neither α nor β characteristics. Eur. J. Pharmacol. *60:* 221–227 (1979).

Mobley, P.L.; Sulser, F.: Adrenal corticoids regulate sensitivity of noradrenaline receptor coupled adenylate cyclase in brain. Nature, Lond. *286:* 608–609 (1980).

Moyer, J.A.; Greenberg, L.H.; Frazer, A.; Brunswick, D.J.; Mendels, J.; Weiss, B.: Opposite effects of acute and repeated administration of desmethylimipramine on adrenergic responsiveness in rat pineal gland. Life Sci. *24:* 2237–2244 (1979).

Olpe, H.R.; Schellenberg, A.: Reduced sensitivity of neurons to noradrenaline after chronic treatment with antidepressant drugs. Eur. J. Pharmacol. *63:* 7–13 (1980).

Pandey, G.N.; Heinse, W.J.; Brown, B.D.; Davis, J.M.: Electroconvulsive shock treatment decreases β-adrenergic receptor sensitivity in rat brain. Nature, Lond. *280:* 234–235 (1979).

Robinson, S.E.; Mobley, P.L.; Smith, H.E.; Sulser, F.: Structural and steric requirements for β-phenethylamines as agonists of the noradrenergic cyclic AMP generating system in the rat limbic forebrain. Arch. Pharmacol. *303:* 175–180 (1978).

Rodbell, M.: The role of hormone receptors and GTP-regulatory proteins in membrane transduction. Nature, Lond. *284:* 17–22 (1980).

Rodbell, M.; Birnbaumer, L.; Pohl, L.L.; Kraus, H.M.: The glucagon sensitive adenylate cyclase system in plasma membranes of rat liver. V. An obligatory role of guanyl nucleotides in glucagon action. J. biol. Chem. *246:* 1877–1882 (1971).

Rodbell, M.; Lin, M.C.; Salomar, Y.: Evidence for interdependent action of glucagon and nucleotides on the hepatic adenylate cyclase system. J. biol. Chem. *249:* 59–65 (1974).

Sarai, K.; Frazer, A.; Brunswick, D.; Mendels, J.: Desmethylimipramine-induced decrease in β-adrenergic receptor binding in rat cerebral cortex. Biochem. Pharmac. *27:* 2179–2181 (1978).

Schildkraut, J.J.: The catecholamine hypothesis of affective disorders. Am. J. Psychiat. *122:* 509–522 (1965).

Schmidt, M.J.; Thornberry, J.F.: Norepinephrine-stimulated cyclic AMP accumulation in brain slices in vitro after serotonin depletion or chronic administration of selective amine reuptake inhibitors. Archs int. Pharmacodyn. Ther. *229:* 42–51 (1977).

Schultz, J.: Psychoactive drug effects on a system which generates cyclic AMP in brain. Nature, Lond. *216:* 417–418 (1970).

Sellinger-Barnett, M.M.; Mendels, J.; Frazer, A.: The effect of psychoactive drugs on beta-adrenergic receptor binding sites in rat brain. Neuropharmacology *19:* 447–454 (1980).

Siggins, G.R.; Schultz, J.E.: Chronic treatment with lithium or desipramine alters discharge frequency and norepinephrine responsiveness of cerebellar Purkinje cells. Proc. natn. Acad. Sci. USA *76:* 5987–5991 (1979).

Sulser, F.: Functional aspects of the norepinephrine receptor coupled adenylate cyclase system in the limbic forebrain and its modification by drugs which precipitate or alleviate depression: molecular approaches to our understanding of affective disorders. Pharmakopyschiat. *11:* 43–52 (1978).

Sulser, F.: New perspectives on the mode of action of antidepressant drugs. Trends pharmacol. Sci. *1:* 92–94 (1979).

Vetulani, J.; Stawarz, R.J.; Sulser, F.: Adaptive mechanisms of the noradrenergic cyclic AMP generating system in the limbic forebrain of the rat: adaptation to persistent changes in the availability of norepinephrine (NE). J. Neurochem. *27:* 661–666 (1976a).

Vetulani, J.; Stawarz, R.J.; Dingell, J.V.; Sulser, F.: A possible common mechanism of action of antidepressant treatments. Arch. Pharmacol. 293: 109–114 (1976b).

Vetulani, J.; Sulser, F.: Action of various antidepressant treatments reduces reactivity of noradrenergic cyclic AMP-generating system in limbic forebrain. Nature, Lond. *257:* 495–496 (1975).

Wagner, H.R.; Crutcher, K.A.; Davis, J.N.: Chronic estrogen treatment decreases β-adrenergic responses in rat cerebral cortex. Brain Res. *171:* 147–151 (1979).

Wang, R.Y.; Aghajanian, G.K.: Enhanced sensitivity of amygdaloid neurones to serotonin and norepinephrine after chronic antidepressant treatment. Commum. Psychopharmacol. *4:* 83–90 (1980).

Wirz-Justice, A.; Kafka, M.S.; Naber, D.; Wehr, T.A.: Circadian rhythms in rat brain alpha and beta-adrenergic receptors are modified by chronic imipramine. Life Sci. *27:* 341–348 (1980).

Wolfe, B.B.; Harden, T.K.; Sporn, J.R.; Molinoff, P.B.: Presynaptic modulation of beta-adrenergic receptors in rat cerebral cortex after treatment with antidepressants. J. Pharmac. exp. Ther. *207:* 446–457 (1978).

F. Sulser, MD, Vanderbilt University School of Medicine and Tennessee Neuropsychiatric Institute, Nashville, TN 37217 (USA)

Adv. biol. Psychiat., vol. 7, pp. 100–112 (Karger, Basel 1981)

Possible Neuronal Mechanisms Involved in the Production of the Disinhibitory Action of Sulpiride

P. Jenner, C.D. Marsden

University Department of Neurology, Institute of Psychiatry and King's College Hospital Medical School, Denmark Hill, London, England

The development of substituted benzamide drugs by *Justin-Besancon* et al. [1964, 1967, 1974] as centrally active pharmacological tools has played a major role in modern psychopharmacology. Within a relatively simple molecular framework a number of closely related drug molecules with a variety of central actions have been introduced into therapeutic use (table I). In the context of the present volume sulpiride is a compound of considerable interest, since not only is this drug a proven anti-emetic [*Corsini* et al., 1976] and anti-psychotic agent [*Pichot and Dreyfus,* 1977], without demonstrating the sedative properties of the closely related molecule sultopride and other more typical neuroleptic agents [*Aschoff* et al., 1975], but it is also said to exert other therapeutic actions in schizophrenia and depression.

The therapeutic uses of sulpiride apparently are dependent on the dose administered (table II). Thus, at low doses the compound acts, for example, as a potent anti-emetic agent. At an intermediate dose level its beneficial effects in depression are observed. At high doses the drug is effective in the control of the psychotic features of schizophrenia. Perhaps the most remarkable additional property claimed for this compound is its ability to exert a disinhibitory action in the treatment of depression and schizophrenia [*Borenstein* et al., 1969; *Collard,* 1969; *Sutter,* 1971; *Ropert* et al., 1978].

The disinhibitory effects of sulpiride in the clinic are discussed by others in the present volume [see chapters by *Borenstein and Gekiere, Aylward* et al. and *Nissen,* this volume]. In brief, sulpiride has been reported to reverse some of the negative symptoms of schizophrenia thereby improving conversation and social contact. Patients are said to lose the 'devitalised appearance' and there is a 'restoration of psychomotor initiative', but with no evidence of sedation. In clinical trials in depression, sulpiride is reported to cause psychomotor stimulation so as to increase vigilance, elevate mood, enhance enthusiasm, and improve socially-orientated behaviour. Patients claim a state of well-being and even euphoria.

Table I. Properties of substituted benzamide drugs in current clinical use [*Justin-Besancon* et al., 1964, 1967, 1974]

	Anti-emetic	Anti-dyskinetic	Anti psychotic	Sedative	Disinhibitory
Metoclopramide	+	0	?	0	0
Tiapride	+	+	0	0	0
Sulpiride	+	+	+	0	+
Sultopride	+	0	+	+	?

Table II. Sulpiride dosage in clinical use [data supplied by Delagrange Laboratories]

sulpiride mg/day	Effects
150– 600	gastrointestinal actions, anti-emetic, anti-vertigo
300– 600	anti-depressant, disinhibitory
800–2000	anti-psychotic

How much of the disinhibitory actions of sulpiride in schizophrenia is due to reversal of post-psychotic depression is not yet clear.

The object of the present essay is to examine the neuronal action of sulpiride in the brain in relation to its disinhibitory action. We will suggest that perhaps the disinhibitory actions of sulpiride in depression and schizophrenia represent the specificity of action of sulpiride on cerebral dopamine receptors together with its preferential action at presynaptic dopamine receptor sites.

Does Sulpiride Display Properties Associated with Currently Used Antidepressant Agents?

Acute administration of commonly used antidepressant agents either causes a blockade of reuptake of the monoamines noradrenaline, 5-hydroxytryptamine or dopamine (tricyclics and tetracyclics), or inhibition of the monoamine oxidase enzymes responsible for their degradation. Both produce a functional increase in synaptic transmitter content. However, this acute increase in functional monoamine levels disappears on repeated administration, to be replaced by postsynaptic receptor subsensitivity [*Sulser,* 1978; *Sulser* et al., 1978; *Bergstrom and Kellar,* 1979; *Segawa* et al., 1979; *Olpe and Schellenberg,* 1980], which is now held to be responsible for the clinical efficacy of these drugs, as is discussed elsewhere in this volume.

Table III. Specificity of sulpiride for cerebral dopamine receptors [data by courtesy from *Leysen* et al., 1980a, b]

Receptor	Ligand	Brain area	Ki (n*M*) values for drug displacement				
			sulpiride	haloperidol	pimozide	trifluo-perazine	thioridazine
Dopamine	^{3}H-spiperone	striatum	51	1.2	0.9	3.9	13
5HT	^{3}H-spiperone	frontal cortex	>10,000	48	211	41	36
H_1-Histamine	^{3}H-mepyramine	cerebellum	>10,000	4,390	>10,000	51	41
α_1-Adrenergic	^{3}H WB-4101	forebrain	> 1,000	8.1	41	20	3.2
α_2-Adrenergic	^{3}H-clonidine	cortex	>10,000	>10,000	>10,000	>10,000	1,383
Acetycholine	^{3}H-dexetimide	striatum	>10,000	4,370	1,022	>10,000	78

Does sulpiride have the same effects? Indirect evidence suggests that such actions are unlikely for sulpiride does not alter 5HT or NA utilisation or turnover, suggesting no alteration of 5HT or noradrenergic synaptic activity [*Fuxe* et al.; 1977 unpubl. data]. In addition, sulpiride does not block myoclonic activity in guinea-pigs induced by the administration of the 5HT precursors *L*-5-hydroxytryptophan, indicating no direct 5HT receptor-blocking properties [unpubl. data]. Further, there is no data to suggest that sulpiride alters monoamine reuptake or monoamine oxidase activity, but surprisingly, direct evidence on these points is lacking. Certainly from our own studies, sulpiride does not alter striatal dopamine reuptake [unpubl. data], but apparently its effects on the uptake of noradrenaline and 5HT, and on monoamine oxidase have not been reported. As to effects on receptors, sulpiride shows little ability to interact with postsynaptic 5HT or noradrenaline receptors (see later).

On present evidence there is little reason to believe that sulpiride produces disinhibitory effects by a mechanism related to that of typical antidepressants.

Receptor Specificity of Sulpiride in Brain

The development of ligand binding techniques has made it possible to obtain receptor profiles of drugs utilising in vitro techniques. A comparison of the receptor binding profiles of a variety of neuroleptic drugs reveals that sulpiride has a remarkable specificity of action (table III). Thus, although sulpiride causes only a relatively weak displacement of ^{3}H-spiperone from striatal dopamine receptors, it has virtually no ability to interact at noradrenaline, 5HT, acetylcholine or histamine receptors [*Leysen* et al., 1980a, b]. In addition, sulpiride is without effect on specific GABA receptor binding [*Hyttel,* 1979]. In contrast, haloperidol or pimozide, although more potent than sulpiride in interacting with dopamine receptors, also exert activity at 5HT and α_1-adrenergic receptors. Similarly, the phenothiazine derivatives trifluoperazine and thioridazine are more active than sulpiride at dopamine receptors, but these compounds can also interact with 5HT, H_1-histamine and α_1-adrenergic receptors, and in the case of thioridazine with muscarinic receptors.

Interaction of Sulpiride with Cerebral Dopamine Mechanisms

As might be expected for a neuroleptic compound acting specifically at cerebral dopamine receptors sulpiride induces the behavioural, electrophysiological and biochemical changes characteristic of a dopamine antagonist. However, sulpiride differs from more typical neuroleptic agents in a number of respects suggesting it may act specifically on a subpopulation of cerebral dopamine receptor sites [*Jenner and Marsden,* 1979a–c].

Behavioural Effects of Sulpiride

In man, sulpiride blocks the emetic effects of apomorphine. Although sulpiride in low doeses does not prevent the beneficial actions of apomorphine in Parkinson's disease [*Corsini* et al., 1976], in higher doses it does counteract the therapeutic effects of *L*-dopa [*Schachter* et al., 1980]. Sulpiride also suppresses neuroleptic-induced tardive dyskinesia, and increases drug-induced Parkinsonism [*Casey* et al., 1979]. These actions of sulpiride are attributable to a blockade of post-synaptic dopamine receptors as are a number of its behavioural effects in animals.

Sulpiride blocks climbing behaviour in mice induced by apomorphine [*Protais* et al., 1976; *Puech* et al., 1976; *Costall* et al., 1978], and the hyperactivity caused by bilateral injection of dopamine into the nucleus accumbens of the rat [*Costall and Naylor,* 1976a, b]. Circling behaviour induced by apomorphine or amphetamine in mice with unilateral 6-hydroxydopamine lesions of cerebral dopamine pathways is inhibited by sulpiride [*Elliott* et al., 1977] (although in rats, while inhibition of the circling response to amphetamine has been reported [*Fuxe* et al., 1976], we and others have been unable to inhibit apomorphine-induced rotation [unpubl. data; *Honda* et al., 1977]).

Despite this apparently typical spectrum of activity for a dopamine antagonist, a number of other behaviours mediated by dopamine are not influenced by sulpiride. Thus, sulpiride in rats does not induce pronounced catalepsy [*Costall and Naylor,* 1975; *Elliott* et al., 1977; *Jenner* et al., 1978], it does not block apomorphine or amphetamine-induced stereotyped behaviour particularly licking, gnawing or biting [*Costall and Naylor,* 1975; *Puech* et al., 1976; *Honda* et al., 1977; *Olpe* 1978; *Jenner* et al., 1978], and it does not antagonise apomorphine-induced behavioural changes in marmosets (checking, locomotion, inactivity) [*Ridley* et al., 1980]. Indeed, the overall effects of sulpiride in blocking post-synaptic actions of dopamine agonists are relatively weak compared to conventional neuroleptic agents.

Biochemical and Electrophysiological Effects of Sulpiride

In man sulpiride causes pronounced elevations of circulating prolactin levels [*Mancini* et al., 1976; *Muller* et al., 1979] and of CSF homovanillic acid concentrations (without altering 5HIAA or MOPEG) [*Bjerkenstedt* et al., 1979], again suggesting an action on cerebral dopamine receptors.

A direct effect of sulpiride at cerebral dopamine receptors has been demonstrated in rat substantia nigra zona compacta by the ability of sulpiride to block the inhibitory effects of dopamine on nigral cell firing [*Pinnock* et al., 1979]. As might be expected, the peripheral administration of sulpiride to rodents is accompanied by an increase in dopamine turnover in striatal, mesolimbic and mesocortical brain areas, as demonstrated by elevation of the dopamine metabolites HVA and DOPAC, and by the altered levels of dopamine following inhibition of synthesis or degradation [*Restelli* et al., 1975; *Tagliamonte* et al., 1975; *Bartholini,* 1976; *Waldmeier and Maitre,* 1976; *Elliott* et al., 1977; *Scatton* et al., 1977]. We have been unable to demonstrate a differential effect on dopamine turnover in different brain areas, although others have claimed a preferential effect of sulpiride on mesolimbic dopamine turnover. A recent study of the regional displacement of ^{3}H-spiperone binding in vivo by sulpiride indicates a preferential action in the olfactory tubercle, hippocampus, septum and substantia nigra compared to the nucleus accumbens, striatum and frontal cortex [*Kohler* et al., 1979].

As already mentioned, sulpiride displaces ligands such as ^{3}H-haloperidol or ^{3}H-spiperone from their specific binding sites in tissue preparations from dopamine-containing brain areas [*Jenner* et al., 1978]. The specificity of this effect is suggested by the greater ability of the pharmacologically active (–)-isomer to cause displacement compared to the less active (+)-form [*Jenner* et al., 1980]. Some selectivity of action of sulpiride on cerebral dopamine receptors is suggested by the failure of sulpiride (and other benzamides) to displace ^{3}H-cis-flupenthixol from its striatal binding site [*Hyttel,* 1980]. Other classes of neuroleptics do so readily.

The difference between sulpiride and other neuroleptic agents also is demonstrated by the failure of sulpiride to inhibit the dopamine stimulation of striatal, mesolimbic or mesocortical adenylate cyclase activity [*Elliott* et al., 1977; *Scatton* et al., 1977; *Roufogalis* et al., 1976], or to increase cerebral cyclic AMP levels following in vivo drug administration [*Trabucchi* et al., 1975].

Selective Interaction of Sulpiride with a Subpopulation of Dopamine Receptors

The atypical properties of sulpiride and in particular its failure to inhibit the dopamine stimulation of adenylate cyclase has formed a major portion of the evidence for the currently fashionable subdivision of cerebral dopamine receptors. It was initially felt that receptors may be divided into those linked to adenylate cyclase and those acting independently of this enzyme [*Kebabian and Calne,* 1979]. However, there is now some doubt as to whether the receptor identified by radioactive ligands is closely linked to adenylate cyclase. There are nevertheless numerous other reasons for the subdivision of cerebral dopamine receptors [*Jenner and Marsden,* 1980].

By the use of ^{3}H-sulpiride we have demonstrated a specific striatal neuronal binding site for this ligand [*Theodorou* et al., 1979]. Lesion studies have demonstrated that these receptor sites are to be found both on the terminals of cortico-

Table IV. Displacement of ^{3}H-sulpiride (15 nM) and ^{3}H-spiperone (0.5 nM) from their striatal binding sites in rat tissue by a range of substituted benzamide drug neuroleptics and transmitter substances

Drug	IC_{50}, nM	
	^{3}H-sulpiride	^{3}H-spiperone
(–)-Sulpiride	9	360
(+)-Sulpiride	355	2,000
(–)-Sultopride	3	66
(+)-Sultopride	370	>1,000
Metoclopramide	57	660
Tiapride	135	1,600
(+)-Butaclamol	2	35
(–)-Butaclamol	3,800	>1,000
Cis-flupenthixol	5	3
Trans-flupenthixol	21	3
Haloperidol	7	6
Trifluoperazine	250	22
Thioridazine	48	120
Clozapine	750	300
Dopamine	1,200	3,700
Noradrenaline	>5,000	>10,000
5-Hydroxytryptamine	>5,000	>10,000

striate glutamate fibres and on the striatal cell bodies [*Theodorou* et al., 1980a]. The ligand is displaced by dopamine but not other neurotransmitters and by both other substituted benzamide drugs and typical neuroleptics (table IV). Although in many respects the ligand binds to a site with properties identical to those of ^{3}H-spiperone it differs in one important respect. Unlike the binding of ^{3}H-spiperone, ^{3}H-sulpiride will not interact with its neuronal binding site unless physiological concentrations of sodium ions are included into the incubation medium [*Theodorou* et al., 1980b]. This suggests that either the ligand interacts at a receptor site different to that acted on by typical neuroleptic ligands, or that it binds to a distinct but different portion of the same receptor complex.

Evidence for a Preferential Presynaptic Action of Sulpiride on Dopamine Neurones

Until now we have concentrated on the interaction of sulpiride with post-synaptic dopamine receptors. Dopamine receptors exist also on both the presynaptic terminals of nigrostriatal dopamine neurones and on the dopamine-containing cell bodies in substantia nigra [*Marsden and Jenner,* 1980]. Earlier we suggested that sulpiride in critical dosage may preferentially interact with presynaptic receptors, thereby modulating dopamine neuronal acitivity without alteration of post-synaptic receptor function [*Elliott* et al., 1977].

The administration of low doses of apomorphine to rodents causes sedation (rather than locomotor stimulation), due, it is believed, to a preferential interaction with presynaptic dopamine receptors causing a decrease in dopamine release [*Di Chiara* et al., 1976]. This sedation can be reversed by doses of sulpiride which, in their own right, do not cause inhibition of the post-synaptic action of apomorphine. Similarly, the effect of bilateral adminstration of amphetamine into nucleus accumbens to cause locomotor hyperactivity can be reversed by the bilateral injection of apomorphine into the same site [*Costall* et al., 1980]. This effect can in turn be reversed by the peripheral administration of low doses of sulpiride. The dose of (–)-sulpiride antagonising this effect of apomorphine was one-fifth of that required to block the amphetamine response at the post-synaptic receptor, suggesting a considerable preference of sulpiride to interact with the presynaptic dopamine receptor. Interestingly, no differential effect could be demonstrated for clozapine or thioridazine in this model, and only an approximately twofold selectivity could be shown for pimozide and haloperidol. This would suggest that in low doses sulpiride will show considerably greater selectivity for presynaptic receptors than more typical neuroleptics. Indeed, this preferential effect can also be seen in the marked ability of sulpiride to increase striatal and mesolimbic dopamine turnover (as measured by HVA and DOPAC levels) at doses which have no effect on the behavioural effects of apomorphine at post-synaptic dopamine receptors [unpubl. data]. This may represent a direct effect of tyrosine hydroxylase activity since sulpiride and other benzamides cause a marked reversal of the inhibitory action of apomorphine on the enzyme in rat striatal synaptosome preparations comparable to that of haloperidol [*Mishra,* 1978]. This would suggest that sulpiride at doses acting preferentially at presynaptic dopamine receptors might cause behavioural activation in its own right. Indeed, although definitive evidence is lacking, sulpiride has been reported to enhance the motor stimulation caused by administration of high doses of apomorphine [*Gessa,* pers. commun.].

Possible Mechanisms for the Disinhibitory Effects of Sulpiride

Based on the present evidence of a specific effect of sulpiride on cerebral dopamine systems we will attempt to put forward mechanisms to explain the disinhibitory effect of this drug. We propose that the antipsychotic actions and the disinhibitory effects of sulpiride in depression and schizophrenia may reflect the selectivity of drug action.

In depression, sulpiride is administered in doses lower than those given to patients with schizophrenia. We suggest that these cause a preferential presynaptic action on cerebral dopamine neurones leading to a functional increase in dopamine transmission which results in behavioural and generalised motor and mental arousal.

We associate the antipsychotic actions of large doses of sulpiride in schizophrenia with a blockade of post-synaptic dopamine receptors. However, this effect of the drug is weak and not typical of that produced by classical neuroleptic agents. The disinhibitory effect of sulpiride seen in the period following an acute psychotic episode may therefore result from an absence of psychomotor retardation and sedation due to the specific actions of the drug on post-synaptic dopamine receptors combined with an antidepressant action of the drug due to its strong presynaptic effects. It can be argued that since the antipsychotic action of sulpiride is due to a blockade of post-synaptic receptors, the presynaptic effects of the drug would no longer influence synaptic transmission. This might suggest that the fact that the drug causes little sedation or motor retardation may be of primary importance. However, since we believe sulpiride to act on a limited population of post-synaptic dopamine receptors it can be envisaged that it may exert a presynaptic effect on all presynaptic terminals but only cause blockade of some post-synaptic receptors. Under these circumstances both disinhibitory and antipsychotic actions of sulpiride could be manifest at the same time. Alternatively there is some evidence to suggest an inverse relationship between dopamine post-synaptic receptor subpopulations [*Calne,* pers. commun.]. An action of sulpiride causing blockade at one type of receptor site might therefore cause activation of other sites. Again this might explain the simultaneous occurrence of both antipsychotic and disinhibitory drug action. In support of this idea we have recently demonstrated that sulpiride stimulates basal striatal adenylate cyclase activity while having no inhibitory action on the dopamine-induced stimulation [unpubl. data].

Summary

In summary, sulpiride apparently exerts a disinhibitory effect in both depression and schizophrenia, but this does not appear to be due to a mechanism associated with typical antidepressants. Sulpiride acts selectively as a dopamine antagonists and may interact preferentially with a subpopulation of presynaptic dopamine receptors. We propose that the disinhibitory effects of sulpiride in depression and schizophrenia may be due to this preferential presynaptic action of the drug on dopamine neurones.

Acknowledgements

This work was supported by the Wellcome Trust, the Medical Research Council, and the Research Funds of the Bethlem Royal and Maudsley Hospitals and King's College Hospital. We are grateful to SESIF (France) for generous supplies of sulpiride. We thank Dr. *Josee Leysen* for allowing us to use the neuroleptic receptor profile data shown in table III.

References

Aschoff, J.C.; Becker, W.; Weinert, D.: Computer analysis of eye movements: evaluation of the state of alertness and vigilance after sulpiride medication. J. Pharm. clin. *2:* 93–97 (1975).

Bartholini, G.: Differential effect of neuroleptic drugs on dopamine turnover in the extrapyramidal and limbic system. J. Pharm. Pharmac. *28:* 429–433 (1976).

Bergstrom, D.A.; Kellar, K.J.: Adrenergic and serotonergic receptor binding in rat brain after chronic desmethylimipramine treatment. J. Pharmac. exp. Ther. *209:* 256–261 (1979).

Bjerkenstedt, L.; Harnryd, C.; Sedvall, G.: The effect of sulpiride on monoaminergic mechanisms in psychotic women. Psychopharmacology *64:* 135–139 (1979).

Borenstein, P.; Champion, C.; Cujo, P.; Gekiers, P.; Olivenstein, C.; Kramarz, P.: An original psychotropic drug: sulpiride. Sem. Hôp. Paris *45:* 1301–1314 (1969).

Casey, D.E.; Gerlach, J.; Simmelsgaard, H.: Sulpiride in tardive dyskinesia. Psychopharmacology *66:* 73–77 (1979).

Collard, J.: Sulpiride, a unique anti-autistic and thymoanaleptic drug. Sem. Hôp. Paris *45:* 3028–3033 (1969).

Corsini, G.U.; Del Zompo, M.; Cianchetti, C.; Mangoni, A.; Gessa, G.L.: Therapeutic efficacy of a combination of apomorphine with sulpiride or metoclopramide in parkinsonism. Psychopharmacology *47:* 169–173 (1976).

Costall, B.; Fortune, D.H.; Hui, S-C.G.; Naylor, R.J.: Neuroleptic antagonism of the motor inhibiting effects of apomorphine within the nucleus accumbens; drug interactions at presynaptic receptors? Eur. J. Pharmacol. *63:* 347–358 (1980).

Costall, B.; Naylor, R.J.: Detection of the neuroleptic properties of clozapine, sulpiride and thioridazine. Psychopharmacologia *43:* 69–74 (1975).

Costall, B.; Naylor, R.J.: A comparison of the abilities of typical neuroleptic agents and of thioridazine, clozapine, sulpiride and metoclopramide to antagonise the hyperactivity induced by dopamine applied intracerebrally to areas of the extrapyramidal and mesolimbic systems. Eur. J. Pharmacol. *40:* 9–19 (1976a).

Costall, B.; Naylor, R.J.: Antagonism of the hyperactivity induced by dopamine applied intracerebrally to the nucleus accumbens septi by typical neuroleptics and by clozapine, sulpiride and thioridazine. Eur. J. Pharmacol. *35:* 161–168 (1976b).

Costall, B.; Naylor, R.J.; Nohria, V.: Climbing behaviour induced by apomorphine in mice: a potential model for the detecting of neuroleptic activity. Eur. J. Pharmacol. *50:* 39–50 (1978).

Di Chiara, G.; Porceddu, M.L.; Vargiu, L.; Argiolas, A.; Gessa, G.L.: Evidence for dopamine receptors mediating sedation in the mouse brain. Nature, Lond. *264:* 564–567 (1976).

Elliott, P.N.C.; Jenner, P.; Huizing, G.; Marsden, C.D.; Miller, R.: Substituted benzamides as cerebral dopamine antagonists in rodents. Neuropharmacology *16:* 333–342 (1977).

Fuxe, K.; Ogren, S-O.; Fredholm, B.; Agnati, L.; Hokfelt, T.; Perez de la Mora, M.: Possibilities of a differential blockade of central monoamine receptors; in Proc. Vth Bel-Air Symp., Geneve 1976.

Honda, F.; Satoh, Y.; Shimamura, K.; Satoh, H.; Noguchi, H.; Uchida, S.; Kato, R.: Dopamine receptor blocking activity of sulpiride in the central nervous systm. Jap. J. Pharmacol. *27:* 397–411 (1977).

Hyttel, J.: Characterization of ^{3}H-GABA receptor binding to rat brain synaptosomal membranes: effect of non-GABAergic compounds. Pyschopharmacology *65:* 211–214 (1979).

Hyttel, J.: Further evidence that ^{3}H-cis(Z)-flupenthixol binds to the adenylate cyclase-associated dopamine receptor (D-1) in rat corpus striatum. Psychopharmacology *67:* 107–109 (1980).

Jenner, P.; Clow, A.; Reavill, C.; Theodorou, A.; Marsden, C.D.: Stereoselective actions of substituted benzamide drugs on cerebral dopamine mechanisms. J. Pharm. Pharmac. *32:* 39–44 (1980).

Jenner, P.; Elliott, P.N.C.; Clow, A.; Reavill, C.; Marsden, C.D.: A comparison of in vitro and in vivo dopamine receptor antagonism produced by substituted benzamide drugs. J. Pharm. Pharmac. *32:* 46–48 (1978).

Jenner, P.; Marsden, C.D.: The substituted benzamides – a novel class of dopamine antagonists. Life Sci. *25:* 479–486 (1979a).

Jenner, P.; Marsden, C.D.: The mechanism of action of substituted benzamide drugs; in Spano, Trabucchi, Corsini, Gessa, Sulpiride and other benzamides, pp. 119–147 (Italian Brain Research Foundation Press, Milan 1979).

Jenner, P.; Marsden, C.D.: Mechanisme d'action des benzamides substitués. Psychol. méd. *11:* 2775–2793 (1979c).

Jenner, P.; Marsden, C.D.: Multiple dopamine receptors in brain; in Clifford-Rose, Metabolic disorders of the nervous system (in press, 1980).

Justin-Besançon, L.; Laville, C.; Thominet, K.: Le métoclopramide et ses homologues. Introduction à leur étude biologique. C.r. hebd. Séanc. Acad. Sci., Paris *258:* 4384–4386 (1964).

Justin-Besançon, L.; Laville, C.; Margarit, J.; Thominet, M.: Constitution chimique et propriétés biologiques d'*o*-anisamines substituées à function alkylsulfone. C.r. hebd. Séanc. Acad. Sci., Paris *279:* 375–376 (1974).

Justin-Besançon, L.; Thominet, M.; Laville, C.; Margarit, J.: Constitution chimique et propriétés biologiques du sulpiride. C.r.hebd. Séanc. Acad. Sci., Paris *265:* 1253–1254 (1967).

Kebabian, J.W.; Calne, D.B.: Multiple receptors for dopamine. Nature, Lond. *277:* 93–96 (1979).

Kohler, C.; Ogren, S-O.; Haglund, L.; Angeby, T.: Regional displacement by sulpiride of ^{3}H-spiperone binding in vivo. Biochemical and behavioural evidence for a preferential action on limbic and nigral dopamine receptors. Neurosci. Lett. *13:* 51–56 (1979).

Leysen, J.E.; Laduron, P.M.; Niemegeers, C.J.E.: Receptor binding profiles, pharmacological activities and clinical properties of drugs. 12th CINP Congr., Gothenburg 1980a.

Leysen, J.E.; Laduron, P.M.; Niemegeers, C.J.E.: Receptor binding profiles, pharmacological activities and clinical properties of drugs. Int. Pharmacopsychiat. (in press, 1980b).

Mancini, A.M.; Guitelman, A.; Vargas, C.A.; Debeljuk, L.; Aparicio, W.J.: Effect of sulpiride on serum prolactin levels in humans. J. clin. Endocr. Metab. *42:* 181–184 (1976).

Marsden, C.D.; Jenner, P.: The pathophysiology of extrapyramidal side-effects of neuroleptic drugs. Psychol. Med. *10:* 55–72 (1980).

Mishra, R.K.: Effect of substituted benzamide drugs on rat striatal tyrosine hydroxylase. Eur. J. Pharmacol. *51:* 189–190 (1978).

Muller, E.E.; Stefanini, E.; Camanni, F.; Locatelli, V.; Massara, F.; Spano, P.F.; Cocchi, D.: Prolaction releasing effect of sulpiride isomers in rats and man. J. neural. Transm. *46:* 205–214 (1979).

Olpe, H-R.: Pharmacological manipulations of the automatically recorded biting behaviour evoked in rats by apomorphine. Eur. J. Pharmacol. *51:* 441–448 (1978).

Olpe, H-R.; Schellenberg, A.; Reduced sensitivity of neurons to noradrenaline after chronic treatment with antidepressant drugs. Eur. J. Pharmacol. *63:* 7–13 (1980).

Pichot, P.; Dreyfus, J.F.: A multicentre double-blind clinical comparison of chlorpromazine and sulpiride in acute psychoses. VIth Wld Congr. Psychiat., Wld psychiat. Ass., Honolulu 1977.

Pinnock, R.D.; Woodruff, G.N.; Turnbull, M.J.: Sulpiride blocks the action of dopamine in the rat substantia nigra. Eur. J. Pharmacol. *56:* 413–414 (1979).

Protais, P.; Constentin, J.; Schwartz, J.C.: Climbing behaviour induced by apomorphine in mice: a simple test for the study of dopamine receptors in striatum. Psychopharmacology *50:* 1–6 (1976).

Puech, A.J.; Simon, P.; Boissier, J-R.: Antagonism by sulpiride of three apomorphine-induced effects in rodents. Eur. J. Pharmacol. *36:* 439–441 (1976).

Restelli, A.; Lucchini, D.; Glasser, A.: Stimulation by sulpiride, chlorpromazine and haloperidol of dopamine turnover in the cerebral cortex of the rat. Pharmacol. Res. Commum. *7:* 409–418 (1975).

Ridley, R.M.; Scraggs, P.R.; Baker, H.F.: The effects of metoclopramide, sulpiride and the stereoisomers of baclofen on amphetamine-induced behaviour in the marmoset. Biol. Psychiat. *15:* 265–274 (1980).

Ropert, R.; Caillard, F.; Petitjean, F.: Benzamides and related compounds. Jt Meet. American Psychiatric Association and the Societe Medico-Psychologique, Atlanta 1978.

Roufogalis, B.D.; Thornton, M.; Wade, D.N.: Specificity of the dopamine sensitive adenylate cyclase for antipsychotic antagonists. Life Sci. *19:* 927–934 (1976).

Scatton, B.; Bischoff, S.; Dedek, J.; Korf, J.: Regional effects of neuroleptics on dopamine metabolism and dopamine sensitive adenylate cyclase. Eur. J. Pharmacol. *44:* 287–292 (1977).

Schachter, M.; Bedard, P.; Debono, A.G.; Jenner, P.; Marsden, C.D.; Price, P.; Parkes, J.D.; Keenan, J.; Smith, B.; Rosenthaler, J.; Horowski, R.; Dorow, R.: The role of D-1 and D-2 receptors. Nature, Lond. *286:* 157–159 (1980).

Segawa, T.; Mizuta, T.; Nomura, Y.: Modifications of central 5-hydroxytryptamine binding sites in synaptic membranes from long-term administration of tricyclic antidepressants. Eur. J. Pharmacol. *58:* 75–83 (1979).

Sulser, F.: Functional aspects of the norepinephrine receptor coupled adenylate cyclase system in the limbic forebrain and its modification by drugs which precipitate or alleviate depression: molecular approaches to an understanding of effective disorders. Pharmakopsychiat. *11:* 43–52 (1978).

Sulser, F.; Vetulani, J.; Mobley, P L. Mode of action of antidepressant drugs. Biochem. Pharmac. *27:* 257–261 (1978).

Sutter, J.M.: Controlled trial of sulpiride in psychiatry. Sem. Hôp. Paris *47:* 446–455 (1971).

Tagliamonte, A.; De Montis, G.; Olianas, M.; Vargiu, L.; Corsini, G.U.; Gessa, G.L.: Selective increase of brain dopamine synthesis by sulpiride. J. Neurochem. *24:* 707–710 (1975).

Theodorou, A.; Crockett, M.; Jenner, P.; Marsden, C.D.: Specific binding of ^{3}H-sulpiride to rat striatal preparations. J. Pharm. Pharmac. *31:* 424–426 (1979).

Theodorou, A.; Hall, M.D.; Jenner, P.; Marsden, C.D.: Cation regulation differentiates specific binding of ^{3}H-sulpiride and ^{3}H-spiperone to rat striatal preparations. J. Pharm. Pharmac. *32:* 441–444 (1980a).

Theodorou, A.; Reavill, C.; Jenner, P.; Marsden, C.D.: Kainic acid lesions of striatum and decortication reduce specific ^{3}H-sulpiride binding in rats, so D-2 receptors exist post-synaptically on cortico-striate afferents and striatal neurons. Life Sci. (in press, 1980b).

Trabucchi, M.; Longoni, R.; Fresia, P.; Spano, P.F.: Sulpiride: a study of the effects on dopamine receptors in rat neostriatum and limbic forebrain. Life Sci. *17:* 1551–1556 (1975).

Waldmeier, P.C.; Maitre, L.: On the relevance of preferential increases of mesolimbic versus striatal dopamine turnover for the production of antipsychotic activity of psychotropic drugs. J. Neurochem. *27:* 589–597 (1976).

P. Jenner, PhD, University Department of Neurology, Institute of Psychiatry and King's College Hospital Medical School, Denmark Hill, London SE5 8AF (England)

Adv. biol. Psychiat., vol. 7, pp. 113–125 (Karger, Basel 1981)

The Potential for the Use of Tryptophan and Other Amine Precursors in the Treatment of Affective Disorders

Simon N. Young[1]

Department of Psychiatry, McGill University, Montreal, Que., Canada

Introduction

There are two main reasons for studying the action of amine precursors in patients with affective disorders. First, compared with normal antidepressant drugs their biochemical actions are relatively specific. Thus, by comparing the actions of different amine precursors it should be possible to obtain some information on which neuronal systems can be manipulated to produce therapeutically useful behavioral and mood changes. Second, because the amine precursors are natural products, and some of them are dietary components, they usually have low toxicity and produce few, if any, side effects. Therefore, this line of research may lead to psychopharmacological agents which are not only of theoretical interest but, because of their lack of side effects, are useful also.

This paper deals with the behavioral and biochemical effects (established, putative and speculative) of six different amine precursors. They are the indoleamine precursors *L*-tryptophan and *L*-5-hydroxytryptophan (5HTP) and also dopa, tyrosine and the *D* and *L* isomers of phenylalanine. The latter group of compounds are precursors of the catecholamines and/or structurally related trace amines. All these compounds, with the exception of *L*-phenylalanine, have been or are being tested in depressed patients and tryptophan has also been tested in mania. I sometimes feel that the more data there is on any of these compounds the less agreement there is on its behavioral effects. This may partly reflect the fact that these compounds are natural products and, therefore, the human body has regulatory mechanisms to control their concentrations and actions. They must be given in ways that take into account these homeostatic mechanisms. For example, they are all metabolized faster than most antidepressants and, therefore, should not be given as a single daily dose. Not all the stud-

[1] The author's work is supported by the Medical Research Council of Canada.

ies have taken into account such simple considerations. Thus, it is important when studying the action of amine precursors to consider their peripheral and central metabolism as well as their behavioral effects.

Indoleamine Precursors

The Antidepressant Actions of Tryptophan and 5HTP

The original studies on tryptophan tested the ability of this amino acid to potentiate the antidepressant effect of monoamine oxidase inhibitors. Both *Coppen* et al. [1963] and *Pare* [1963] found a significant potentiation and this was subsequently confirmed by other workers [*Glassman and Platman,* 1969; *Ayuso Gutierrez and Lopez-Ibor Alino,* 1971]. *Coppen* et al. [1967] went on to study the action of tryptophan alone and, because they obtained favorable results, their report was followed up in many other studies. These have been reviewed several times recently [*Mendels* et al., 1975; *Wirz-Justice,* 1977; *d'Elia* et al., 1978; *Cooper,* 1979]. The results with tryptophan alone or in combination with tricyclic antidepressants have been very variable and there is no consensus even among the reviewers as to whether it is an effective antidepressant.

The results with 5HTP have been more consistant. It does seem to be an antidepressant when given at an adequate dose [*Sano,* 1972; *Takahashi* et al., 1975; *Angst* et al., 1977; *Kaneko* et al., 1979]. *Van Praag* et al. [1972] found that patients with low CSF 5-hydroxyindoleacetic acid (suggesting low brain 5HT metabolism) responded better than others to 5THP, a finding confirmed by *Fujiwara and Otsuki* [1974] but not by *Takahashi* et al. [1975]. In a further series of elegant experiments *van Praag* went on to show that 5HTP exhibits a prophylactic effect in patients with recurrant depression and that patients who respond best to the prophylactic effect are also those with low CSF 5-hydroxyindoleacetic acid [*van Praag,* 1979, 1980; *van Praag and de Haan,* 1979]. 5HTP is also capable of potentiating the antidepressant action of the tricyclic antidepressant chomipramine [*van Praag* et al., 1974; *van Praag,* 1980] and the monoamine oxidase inhibitor nialamide [*Lopez-Ibor Alino* et al., 1976].

A Comparison of the Biochemical Effects of Tryptophan and 5HTP

If tryptophan and 5HTP differ in their clinical effects, with the antidepressant effect of 5HTP being seen much more clearly than that of tryptophan (if indeed it is an antidepressant), the question arises as to why this should be so if they are both 5HT precursors. Some of the similarities and differences between tryptophan and 5HTP are summarized in table I. One of the most important differences is that 5HTP can produce much larger increases in brain 5HT than tryptophan can [*Moir and Eccleston,* 1968], presumably because tryptophan hydroxylase is closer to saturation with tryptophan than aromatic amino

Table I. A comparison between the effects of tryptophan and 5HTP

	Tryptophan	5HTP
5HT synthesis	↑	↑ ↑
Firing of 5HT neurons	↓	↓
5HT release	↑ ?	↑
Tryptamine synthesis	↑	–
Catecholamine release	–	↑
Brain protein synthesis	↑	↓

acid decarboxylase is with 5HTP. Under normal conditions release of 5HT is closely related to firing of 5HT (raphe) neurons [*Puizillout* et al., 1979]. Both tryptophan and 5HTP inhibit the firing of raphe neurcns due to the action of 5HT on presynaptic 5HT receptors [*Gallager and Aghajanian,* 1976]. In the case of tryptophan this limits the increase in 5HT release, but with 5HTP the increase in 5HT levels can be large enough that there is release of 5HT independent of neuronal firing. This has been referred to as 'spillover' of 5HT onto postsynaptic receptors. *Ternaux* et al. [1976] studied release of 5HT into the ventricles of the rat. They found that acute injection of 5HTP produced an immediate, important and long-lasting increase of 5HT release while acute injection of tryptophan led to a transient and moderate elevation of 5HT release which was detected only during the 2nd h of perfusion. This explains why tryptophan administration has little or no gross behavioral effects in experimental animals [*Modigh,* 1973] while 5HTP can produce hyperactivity [*Modigh,* 1972]. Tryptophan can lead to hyperactivity when it is given with a monoamine oxidase inhibitor [*Hess and Doepfner, 1961*]. In this case the increase in brain 5HT is larger and there is presumably spillover of 5HT onto postsynaptic receptors even though the tryptophan and the monoamine oxidase inhibitor [*Aghajanian* et al., 1970] tend to inhibit firing of raphe neurons. Thus, both 5HTP and the combination of tryptophan and a monoamine oxidase inhibitor are established as effective antidepressant treatments and both lead to substantial increases in 5HT release. With tryptophan alone the increase in 5HT release is less and its antidepressant effect is doubtful.

There are also other differences between tryptophan and 5HTP (table I). For example, tryptophan increases the synthesis of tryptamine while 5HTP does not. The possible significance of this is discussed below (next section). Also unlike tryptophan, 5HTP can be decarboxylated in catecholaminergic neurons. The 5HT formed there can displace catecholamines [*Awazi and Guldberg,* 1978]. The combination of 5HTP and a peripheral decarboxylase inhibitor does seem to have dopamine agonistic properties in humans as shown by hormonal release [*van Praag* et al., 1976]. However, as dopa is not an antidepressant (see below), this probably does not contribute to the antidepressant effect of 5HTP.

Another difference between tryptophan and 5HTP concerns protein synthesis. Studies on experimental animals suggest that tryptophan increases brain protein synthesis [*Jørgensen and Majumdar,* 1976] while 5HTP disaggregates brain polysomes and decreases protein synthesis [*Weiss* et al., 1975]. It is not possible to say whether this occurs in humans or what the clinical significance of this finding might be.

Of all these factors the one that is probably most relevant to the different clinical results obtained with tryptophan and 5HTP is the difference in 5HT release. However, even if release is smaller with tryptophan than with 5HTP this does not mean that tryptophan can not be clinically effective. I feel that the evidence available at the moment suggests that tryptophan can be an effective antidepressant and that much of the variability in the clinical studies with tryptophan can be explained.

Possible Reasons for the Variable Clinical Results with Tryptophan

There are several possible reasons for the variable clinical results obtained with tryptophan. One is the breakdown of tryptophan in the liver by tryptophan pyrrolase, which can remove tryptophan from the circulation rapidly. When humans were given a 5-gram tryptophan load (about 4 to 5 times the normal daily dietary intake) plasma tryptophan was elevated ten-fold by an hour but was back to normal after 8 h [*van Praag* et al., 1973]. Thus, tryptophan must be given in divided doses to keep plasma tryptophan high. When patients are treated chronically with tryptophan some metabolize tryptophan much faster than others. In two studies we have found that the rise of plasma tryptophan in depressed patients on tryptophan was quite variable and that there was a relationship between the rise of plasma tryptophan and the therapeutic response [*Chouinard* et al., 1978, 1979b]. This suggests that in some patients lack of response to tryptophan may be due to a high rate of tryptophan breakdown in the liver which prevents any long-lasting rise of plasma tryptophan. We have tried to overcome this by using nicotinamide, which in experimental animals inhibits breakdown of tryptophan by tryptophan pyrrolase [*Young and Sourkes,* 1974]. However, even when tryptophan is given with nicotinamide the rise of plasma tryptophan is very variable [*Chouinard* et al., 1978, 1979a, b] suggesting that nicotinamide is of little, if any, use.

Peripheral metabolism may be important in another way also. *Møller* et al. [1976] found that patients who responded to tryptophan were those whose ratio of plasma tryptophan to other large neutral amino acids was low. As the other large neutral amino acids compete with tryptophan for entry into the brain the patients who responded were ones who might have been expected to have low brain tryptophan before treatment.

Another factor that seems to be very important in the clinical use of tryptophan is the dose used. Unipolar and bipolar patients differ in the dose required.

We have pointed out that for unipolar patients there seems to be a therapeutic window, i.e., tryptophan is effective at low doses but not at high doses [*Young and Sourkes,* 1977; *Chouinard* et al., 1979b]. The evidence for this is derived from an analysis of studies in the literature. Most of the studies that found a therapeutic effect with unipolar patients used low doses of tryptophan, while those that used high doses usually found no effect. Also in studies where an incremental dosage schedule was used the therapeutic effect faded with time as higher doses were reached. No such effect was seen in studies using a constant dose. One possible explanation of this decline in therapeutic effect at higher doses is that inhibition of firing of 5HT neurons only becomes appreciable at high tryptophan doses. Thus, at low doses there would be increased synthesis of 5HT, little inhibition of firing of 5HT neurons and thus an appreciable increase in 5HT release. At higher tryptophan doses there would still be increased synthesis of 5HT. However, because of inhibition of firing of 5HT neurons there would be little increase in 5HT release. If this explanation is correct then other drugs which inhibit firing of 5HT neurons, such as tricyclic antidepressants [*Sheard* et al., 1972], should move the therapeutic window to lower tryptophan doses. This does seem to be the case. Of the studies that tested the ability of tryptophan to potentiate the action of tricyclic antidepressants only those which used lower doses of tryptophan found a potentiation [*Lopez-Ibor Alino* et al., 1973; *Shaw* et al., 1975; *Walinder* et al., 1976; *Chouinard* et al., 1979b]. Also in our study we found that there was a significant positive relationship between increase in plasma tryptophan and therapeutic response in patients treated with tryptophan alone, but a significant negative relationship in patients given the same dose of tryptophan with imipramine [*Chouinard* et al., 1979b]. We explained this by suggesting that imipramine moved the therapeutic window to lower tryptophan levels and, therefore, higher plasma tryptophan levels were associated with less therapeutic response.

Monoamine oxidase inhibitors also inhibit firing of 5HT neurons [*Aghajanian* et al., 1970]. However, as discussed above, when tryptophan is given with a monoamine oxidase inhibitor release of 5HT is probably not dependent on neuronal firing. Thus, it is not surprising that tryptophan potentiates the action of monoamine oxidase inhibitors at every dose level from low to high [*Coppen* et al., 1963; *Pare,* 1963; *Glassman and Platman,* 1969; *Ayuso Gutierrez and Lopez-Ibor Alino,* 1971].

Because of the different clinical responses of unipolar patients at different doses we have studied the metabolic effects of different dose levels [*Young and Gauthier,* unpublished data]. Neurologic patients about to undergo diagnostic pneumoencephalography were given a small or large tryptophan load. Analysis of the CSF suggested that both small and large loads increased brain 5HT synthesis about two-fold. However, while the small load doubled the rate of tryptamine metabolism in the brain the large load quadrupled it. After the large load the rate of tryptamine metabolism was about 40% of the rate of 5HT metabolism instead of 15% as under basal conditions [*Young* et al., 1980]. Thus, some of the decline in therapeutic effect at high tryptophan doses may be due to enhanced tryptamine metabolism. Perhaps the tryptamine acts preferentially on presynaptic 5HT receptors to enhance the inhibition of firing of 5HT neurons. Studies on the action of tryptamine and 5HT applied iontophoretically to single neurons of rat cortex [*Jones and Boulton,* personal communication] suggest that another mechanism could also be involved. Thus, depressant effects of 5HT were profoundly enhanced by weak concurrent application of tryptamine, which itself did not alter the baseline cell firing rate. However, excitatory responses to 5HT were consistantly reversed into depressant responses during weak applications of tryptamine. Perhaps this could help to account for the reversal of the clinical effect of tryptophan at high tryptophan doses.

The Action of Tryptophan in Depressed and Manic Bipolar Patients

The therapeutic window for tryptophan applies only to unipolar patients, as bipolar depressed patients do seem to show some therapeutic response at high tryptophan doses [*Murphy* et al., 1974; *Farkas* et al., 1976; *Møller* et al., 1976]. The reason for this difference between unipolar and bipolar patients is unknown. Bipolar patients also show response to high doses of tryptophan in the manic phase. Thus, two studies of acutely manic patients showed moderate [*Prange* et al., 1974] or no [*Chambers and Naylor,* 1978] therapeutic effect with 6 g of tryptophan a day, but a third study achieved better results using an average dose of 9.6 g a day [*Murphy* et al., 1974]. We have recently studied the action of 12 g tryptophan a day in acute manic patients [*Chouinard, Young, Annable and Lesage,* unpublished data]. Although at the time of writing the results of this study are not fully analyzed, a preliminary analysis of the data shows that tryptophan was significantly better than placebo in treating the symptoms of mania.

If tryptophan is effective in both depressed and manic phases, it may be useful in the prophylactic treatment of bipolar patients. So far we have obtained encouraging results on the prophylactic use of tryptophan in 1 manic-depressive patient [*Chouinard* et al., 1979a]. These data suggest that tryptophan may have a profile of activity rather like that of lithium. Obviously more studies are needed on the antimanic and prophylactic use of tryptophan and 5HTP in bipolar patients.

Precursors of Catecholamines and Related Compounds

Trials of Dopa in Depressed Patients

There have been relatively few trials of *L*-dopa in depressed patients, but they suggest that it is not an effective antidepressant [for reviews see *Murphy,* 1973; *Mendels* et al., 1975]. When dopa does have some effect it is primarily on motor retardation rather than mood and it may precipitate hypomanic episodes in some bipolar patients.

Metabolic studies show that dopa can increase noradrenaline synthesis as well as dopamine synthesis [*Hornykiewicz,* 1974]. However, the increase in noradrenaline synthesis is probably not appreciable in humans [*Wilk and Mones,* 1971]. Of course, some dopa will be decarboxylated in noradrenergic and 5HT neurons leading to displacement of noradrenaline and 5HT in the same way that 5HTP can cause displacement of catecholamines. However, the effect of dopa is primarily dopaminergic and clinical results suggest that potentiation of dopaminergic function can elevate motor activity but not mood.

Prospects for the Use of Tyrosine in Depressed Patients

Until recently it was usually thought that tyrosine hydroxylase is fully saturated with tyrosine. However, studies on experimental animals suggests that this is not true, although tyrosine hydroxylase is probably closer to saturation with tyrosine than tryptophan hydroxylase is with tryptophan [*Carlsson and Lindqvist,* 1978]. Studies on experimental animals suggest that tyrosine administration can increase noradrenaline synthesis [*Gibson and Wurtman,* 1978] but probably not, under normal circumstances, dopamine synthesis [*Scally* et al., 1977]. In dopaminergic neurons feedback inhibition of tyrosine hydroxylase by dopamine seems to be more important than precursor availability. If tyrosine administration increases noradrenaline synthesis in human brain as in rat brain, it should obviously be tested as an antidepressant. In the only clinical report available so far a single patient showed a good response to tyrosine [*Gelenberg* et al., 1980].

We have studied the behavioral effects of tyrosine in mice using the 'behavioral despair' test of *Porsolt* et al. [1977]. In this test animals are placed in water and, after a while of struggling, they become relatively immobile. Antidepressant drugs, including pharmacologically atypical ones, decrease the immobility [*Porsolt* et al., 1977]. However, this test is not specific for antidepressants [*Browne,* 1979; *Wallach and Hedley,* 1979] and it is relatively insensitive to antidepressant drugs which influence primarily 5HT [*Porsolt* et al., 1979; *Gorka* et al., 1979]. We found that tyrosine, like antidepressants, decreased the immobility of animals in this test [*Gibson, Deikel, Young and Binik,* unpublished data]. This effect of tyrosine may have been partly a stimulant effect as it also increased open field activity.

Trials of D-Phenylalanine in Depressed Patients

The *L* isomer of phenylalanine is readily hydroxylated to tyrosine, but it can also be decarboxylated to phenylethylamine. *D*-Phenylalanine is probably not converted to tyrosine, but it can still be decarboxylated to phenylethylamine. Thus, lower doses of the *D* isomer are needed to raise brain phenylethylamine than of the *L* isomer [*Borison* et al., 1978]. Phenylethylamine is normally present in brain at a concentration much lower than that of the catecholamines, but its rate of synthesis in rat brain is comparable to that of the catecholamines [*Durden and Philips,* 1980]. Phenylethylamine has interesting behavioral properties and the concentration of its metabolite phenylacetic acid is low in the CSF of depressed patients [*Sandler* et al., 1979]. Suggestions have been made that phenylethylamine is involved in the control of mood [*Fischer,* 1975; *Sabelli* et al., 1978] and this explains why *D*-phenylalanine has been tested in depressed patients. *Spatz* et al. [1975] found that *D*-phenylalanine increased urinary phenylethylamine and elevated mood in depressed patients in an open study. *Beckmann* et al. [1977, 1979] tested *DL*-phenylalanine in one open and one double-blind controlled study against imipramine with encouraging results. In these studies the dose of the *L* isomer in the mixture was much less than the normal daily dietary intake of phenylalanine, so any behavioral effects were due to the *D* isomer.

More work is needed on both behavioral and biochemical effects of *D*-phenylalanine. If it is an antidepressant it is important to know how it is working. Although it can be converted to phenylethylamine, there is little other information on how and to what it can be metabolized. Also, phenylethylamine can be converted further to phenylethanolamine [*Saavedra and Axelrod,* 1973], a compound about which little is known. A combination of clinical and experimental studies on *D*-phenylalanine could throw some light on the possible involvement of phenylethylamine in the control of mood.

Prospects for the Use of L-Phenylalanine in Depressed Patients

L-Phenylalanine is the precursor of both tyrosine and phenylethylamine, so its action may be like that of a combination of tyrosine and *D*-phenylalanine. Because the primary metabolic pathway of *L*-phenylalanine is to tyrosine, more of it will be needed to raise the concentration of phenylethylamine than when the *D* isomer is given. Suitable amounts to try in depressed patients would probably be several times the normal daily dietary intake (i.e., at least 6 g per day). So far the only clinical data available are from *Birkmayer* [1966] who noted an elevation of mood when *L*-phenylalanine was infused into parkinsonian patients who were also depressed.

As with tyrosine we have tested the activity of phenylalanine in the 'behavioral despair' test of *Porsolt* et al. [1977]. We found that, in common with antidepressants, *L*-phenylalanine decreased the immobility of the animals [*Gibson, Deikel, Young and Binik,* unpublished data]. It did this without increasing open field activity and thus it looks, in this test, like an antidepressant. There is scope for studies on the action of *L*-phenylalanine in depressed patients and on its ability to increase phenylethylamine and noradrenaline synthesis in human brain.

Conclusions

More work has been done on indoleamine precursors than on the precursors of catecholamines and related compounds. Studies on the latter group of compounds should provide useful information on the theories which implicate noradrenaline and phenylethylamine in the control of mood. The data available at the moment for indoleamine precursors suggests that (i) 5HTP is an antidepressant and also has prophylactic action in depression; (ii) tryptophan potentiates the action of monoamine oxidase inhibitors; (iii) tryptophan is probably an antidepressant when given by itself, but there is a therapeutic window for unipolar patients, and (iv) tryptophan seems to be an antimanic agent.

One implication of these conclusions is that an increase in 5HT release is by itself enough to elevate the mood of some depressed patients. More work is needed on the action of tryptophan and 5HTP in mania and on the prophylactic effects of these amino acids in depression and mania.

The clinical and experimental data on dopa, tyrosine and phenylalanine suggest the following. (i) Dopa, which acts primarily to increase dopamine, is not an antidepressant. (ii) Studies on experimental animals suggest that the main effect of tyrosine is on noradrenaline synthesis. Thus, studies on the action of tyrosine in depressed patients would be a useful test of theories relating noradrenaline to the control of mood. (iv) Initial studies on the antidepressant action of *D*-phenylalanine are encouraging. Further studies are needed to reinvestigate its antidepressant action and to determine whether its behavioral effects (if any) are due to phenylethylamine synthesis. (v) *L*-Phenylalanine would be expected to increase synthesis of noradrenaline and phenylethylamine. Metabolic studies are needed to determine whether this is so for human brain and, if so, what effect it has on depressed patients.

In the long run it may be a combination of these compounds that will be found to be most clinically useful. For example, the combination of *L*-tryptophan and *L*-phenylalanine would be expected to enhance synthesis of 5HT, noradrenaline and phenylethylamine and is unlikely to produce serious side effects. The next step on from using combinations of purified amino acids would be to use dietary manipulations alone. I feel that dietary manipulations are unlikely to produce biochemical changes of sufficient magnitude to be of use in depressed patients, but then some have doubted whether administration of single amino acids could influence mood in therapeutically useful ways.

References

Aghajanian, G.K.; Graham, A.W.; Sheard, M.H.: Serotonin-containing neurons in brain: depression of firing by monoamine oxidase inhibitors. Science *169:* 1100–1101 (1970).

Angst, J.; Woggon, B.; Schoepf, J.: The treatment of depression with *L*-5-hydroxytryptophan versus imipramine: results of two open and one double-blind study. Arch. Psychiat. NervKrankh. *224:* 175–186 (1977).

Ayuso Gutierrez, J.L.; Lopez-Ibor Alino, J.J.: Tryptophan and a MAOI (nialamide) in the treatment of depression: a double-blind study. Int. Pharmacopsychiat. *6:* 92–97 (1971).

Awazi, N.; Guldberg, H.C.: On the interaction of 5-hydroxytryptophan and 5-hydroxytryptamine with dopamine metabolism in the rat striatum. Arch. Pharmacol. *303:* 63–72 (1978).

Beckmann, H.; Athen, D.; Olteanu, M.; Zimmer, R.: *DL*-Phenylalanine versus imipramine: a double-blind controlled study. Arch. Psychiat. NervKrankh. *227:* 49–58 (1979).

Beckmann, H.; Strauss, M.A.; Ludolph, E.: *DL*-Phenylalanine in depressed patients: an open study. J. neural Transm. *41:* 123–134 (1977).

Birkmayer, W.: Experimentelle Befunde und neue Aspekte bei extrapyramidalen Erkrankungen. Wien. Z. NervHeilk. *13:* 128–139 (1966).

Borison, R.L.; Maple, P.J.; Havdala, H.S.; Diamond, B.I.: Metabolism of an amino acid with antidepressant properties. Res. Commun. chem Pathol. Pharmacol. *21:* 363–369 (1978).

Browne, R.G.: Effects of antidepressants and anticholinergics in a mouse 'behavioral despair' test. Eur. J. Pharmacol. *58:* 331–334 (1979).

Carlsson, A.; Lindqvist, M.: Dependence of 5HT and catecholamine synthesis on concentrations of precursor amino acids in rat brain. Arch. Pharmacol. *303:* 157–164 (1978).

Chambers, C.A.; Naylor, G.J.: A controlled trial of *L*-tryptophan in mania. Br. J. Psychiat. *132:* 555–559 (1978).

Chouinard, G.; Jones, B.D.; Young, S.N.; Annable, L.: Potentiation of lithium by tryptophan in a case of bipolar illness. Am. J. Psychiat. *136:* 719–720 (1979a).

Chouinard, G.; Young, S.N.; Annable, L.; Sourkes, T.L.: Tryptophan-nicotinamide, imipramine and their combination in depression. Acta psychiat. scand. *59:* 395–414 (1979b).

Chouinard, G.; Young, S.N.; Annable, L.; Sourkes, T.L.; Kiriakos, R.Z.: Tryptophan-nicotinamide combination in the treatment of newly admitted depressed patients. Commun. Psychopharmacol. *2:* 311–318 (1978).

Cooper, A.J.: Tryptophan antidepressant 'physiological sedative': fact or fancy? Psychopharmacology, Berlin *61:* 97–102 (1979).

Coppen, A.; Shaw, D.M.; Farrell, J.P.: Potentiation of the antidepressant effect of a monoamine-oxidase inhibitor by tryptophan. Lancet *i:* 79–81 (1963).

Coppen, A.; Shaw, D.M.; Herzberg, B.; Maggs, R.: Tryptophan in the treatment of depression. Lancet *ii:* 1178–1180 (1967).

Durden, D.A.; Philips, S.R.: Kinetic measurements of the turnover rates of phenylethylamine and tryptamine in vivo in the rat brain. J. Neurochem. *34:* 1725–1732 (1980).

d'Elia, G.; Hanson, L.; Raotma, H.: *L*-Tryptophan and 5-hydroxytryptophan in the treatment of depression: a review. Acta psychiat. scand. *57:* 239–252 (1978).

Farkas, T.; Dunner, D.L.; Fieve, R.R.: *L*-Tryptophan in depression. Biol. Psychiat. *11:* 295–302 (1976).

Fischer, E.: The phenylethylamine hypothesis of thymic homeostasis. Biol. Psychiat. *10:* 667–673 (1975).

Fujiwara, J.; Otsuki, S.: Subtype of affective psychoses classified by response on amine-precursors and monoamine metabolism. Folia psychiat. neurol. jap. *28:* 93–100 (1974).

Gallager, D.W.; Aghajanian, G.K.: Inhibition of firing of raphe neurons by tryptophan and 5-hydroxytryptophan: blockade by inhibiting serotonin synthesis with Ro-4-46 02. Neuropharmacology *15:* 149–156 (1976).

Gelenberg, A.J.; Wojcik, J.D.; Growdon, J.H.; Sved, A.F.; Wurtman, R.J.: Tyrosine for the treatment of depression. Am. J. Psychiat. *137:* 622–623 (1980).

Gibson, C.J.; Wurtman, R.J.: Physiological control of brain norepinephrine synthesis by brain tyrosine concentration. Life Sci. *22:* 1399–1406 (1978).

Glassman, A.H.; Platman, S.R.: Potentiation of monoamine-oxidase inhibitor by tryptophan. J. psychiat. Res. *7:* 83–88 (1969).

Gorka, Z.; Wojtasik, E.; Kwiatek, H.; Maj, J.: Action of serotoninmimetics in the behavioral despair test in rats. Commun. Psychopharmacol. *3:* 133–136 (1979).

Hess, S.M.; Doepfner, W.: Behavioral effects and brain amine content in rats. Archs int. Pharmacodyn. Thér. *134:* 89–99 (1961).

Hornykiewicz, O.: The mechanisms of action of *L*-dopa in Parkinsons's disease. Life Sci. *15:* 1249–1259 (1974).

Jørgensen, A.J.F.; Majumdar, A.P.N.: Bilateral adrenalectomy: effect of tryptophan force-feeding on amino acid incorporation into ferritin, transferrin and mixed proteins of liver, brain and kidneys in vivo. Biochem. Med. *16:* 37–46 (1976).

Kaneko, M.; Kurnashiro, H.; Takahashi, Y.; Hoshino, Y.: *L*-5-HTP treatment and serum 5-HT level after *L*-5-HTP loading on depressed patients. Neuropsychobiology *5:* 232–240 (1979).

Lopez-Ibor Alino, J.J.; Ayuso Gutierrez, J.L.; Montejo Iglesias, M.L.: Tryptophan and amitriptyline in the treatment of depression: a double-blind study. Int. Pharmacopsychiat. *8:* 145–151 (1973).

Lopez-Ibor Alino, J.J.; Ayuso Gutierrez, J.L.; Montejo Iglesias, M.L.: 5-Hydroxytryptophan (5-HTP) and a MAOI (nialamide) in the treatment of depressions: a double-blind controlled study. Int. Pharmacopsychiat. *11:* 8–15 (1976).

Mendels, J.; Stinnett, J.L.; Burns, D.; Frazer, A.: Amine precursors and depression. Archs gen. Psychiat. *32:* 22–30 (1975).

Modigh, K.: Central and peripheral effects of 5-hydroxytryptophan on motor activity in mice. Psychopharmacologia *23:* 48–54 (1972).

Modigh, K.: Effects of *L*-tryptophan on motor activity in mice. Psychopharmacologia *30:* 123–134 (1973).

Moir, A.T.B.; Eccleston, D.: The effects of precursor loading in the central metabolism of 5-hydroxyindoles. J. Neurochem. *15:* 1108 (1968).

Møller, S.E.; Kirk, L.; Fremming, K.H.: Plasma amino acids as an index for subgroups in main depressive psychosis: correlation to effect of tryptophan. Psychopharmacology, Berlin *49:* 205–213 (1976).

Murphy, D.L.: Mental effects of *L*-dopa. Ann u. Rev. Med. *24:* 209–216 (1973).

Murphy, D.L.; Baker, M.; Goodwin, F.K.; Miller, H.; Kotin, J.; Bunney, W.E.: *L*-Tryptophan in affective disorders: indoleamine changes and differential clinical effects. Psychopharmacologia *34:* 11–20 (1974).

Pare, C.M.B.: Potentiation of monoamine oxidase inhibitors by tryptophan. Lancet *ii:* 527–528 (1963).

Porsolt, R.D.; Bertin, A.; Jalfre, M.: Behavioral despair in mice: a primary screening test for antidepressants. Archs int. Pharmacodyn. Thér. *229:* 327–336 (1977).

Porsolt, R.D.; Bertin, A.; Blavel, N.; Deniel, M.; Jalfre, M.: Immobility induced by forced swimming in rats: effects of agents which modify central catecholamine and serotonin activity. Eur. J. Pharmacol. *57:* 201–210 (1979).

Praag, H.M. van: Central serotonin its relation to depression vulnerability and depression prophylasis, in Obiols, Ballus, Gonzalez Monclus, Pujol, Biological psychiatry today, pp. 485–498 (Elsevier/North-Holland Biomedical Press, Amsterdam 1979).

Praag, H.M. van: Central monoamine metabolism in depressions: serotonin and related compounds. Compren. Psychiat. *21:* 30–43 (1980).

Praag, H.M. van; Burg, W. van den; Bos, E.R.H.; Dols, L.C.W.: 5-Hydroxytryptophan in combination with clomipramine in 'therapy-resistant' depressions. Psychopharmacologia *38:* 267–269 (1974).

Praag, H.M. van; Flentge, F.; Korf, J.; Dols, L.C.W.; Schut, T.: The influence of probenecid on the metabolism of serotonin, dopamine and their precursors in man. Psychopharmacologia *33:* 141–151 (1973).

Praag, H.M. van; Haan, S. de: Central serotonin metabolism and frequency of depression. Psychiat. Res. *1:* 219–224 (1979).

Praag, H.M. van; Korf, J.; Dols, L.C.W.; Schut, T.: A pilot study of the predictive value of the probenecid test in application of 5-hydroxytryptophan as antidepressant. Psychopharmacologia *25:* 14–21 (1972).

Praag, H.M. van; Korf, J.; Lequin, R.M.: An unexpected effect of *L*-5-hydroxytryptophan-ethyl-ester combined with a peripheral decarboxylase inhibitor on human serum prolactin. Psychopharmacol. Commun. *2:* 369–378 (1976).

Prange, A.J.; Wilson, I.C.; Lynn, C.W.; Alltop, L.B.; Stikeleather, R.A.: *L*-Tryptophan in mania: contribution to a permissive hypothesis of affective disorders. Archs gen. Psychiat. *30:* 56–62 (1974).

Puizillout, J.J.; Gaudin-Chazal, G.; Daszuta, A.; Seyfritz, N.; Ternaux, J.P.: Release of endogenous serotonin from 'encephale isole' cats: correlations with raphe neuronal activity and sleep and wakefulness. J. Physiol., Paris *75:* 531–537 (1979).

Saavedra, J.M.; Axelrod, J.: Demonstration and distribution of phenylethanolamine in brain and other tissues. Proc. natn. Acad. Sci. USA *70:* 769–772 (1973).

Sabelli, H.C.; Borison, R.L.; Diamond, B.I.; Havdala, H.S.; Narasimhachari, N.: Phenylethylamine and brain function. Biochem. Pharmac. *27:* 1707–1711 (1978).

Sandler, M.; Ruthven, C.R.J.; Goodwin, B.L.; Coppen, A.: Decreased cerebrospinal fluid concentration of free phenylacetic acid in depressive illness. Clinica chim. Acta *93:* 169–171 (1979).

Sano, I.: *L*-5-Hydroxytryptophan (*L*-5-HTP) therapy. Folia psychiat. neurol. jap. *26:* 7–17 (1972).

Scally, M.C.; Ulus, I.; Wurtman, R.J.: Brain tyrosine level controls striatal dopamine synthesis in haloperidol-treated rats. J. neural Transm. *41:* 1–6 (1977).

Shaw, D.M.; MacSweeney, D.A.; Hewland, R.; Johnson, A.L.: Tricyclic antidepressants and tryptophan in unipolar depression. Psychol. Med. *5:* 276–278 (1975).

Sheard, M.H.; Zolovick, A.; Aghajanian, G.K.: Raphe neurons: effect of tricyclic antidepressant drugs. Brain Res. *43:* 690–694 (1972).

Spatz, H.; Heller, B.; Nachon, M.; Fisher, E.: Effects of *D*-phenylalanine on clinical picture and phenethylaminuria in depression. Biol. Psychiat. *10:* 235–239 (1975).

Takahashi, S.; Kondo, H.; Kato, N.: Effect of *L*-5-hydroxytryptophan on brain monoamine metabolism and evaluation of its clinical effect in depressed patients. J. psychiat. Res. *12:* 177–187 (1975).

Ternaux, J.P.; Boireau, A.; Bourgoin, S.; Hamon, M.; Hery, F.; Glowinski, J.: In vivo release of 5-HT in the lateral ventricle of the rat: effects of 5-hydroxytryptophan and tryptophan. Brain Res. *101:* 533–548 (1976).

Walinder, J.; Skott, A.; Carlsson, A.; Nagy, A.; Roos, B.E.: Potentiation of the antidepressant action of clomipramine by tryptophan. Archs gen. Psychiat. *33:* 1384–1389 (1976).

Wallach, M.B.; Hedley, L.R.: The effects of antihistamines in a modified behavioral despair test. Commun. Psychopharmacol. *3:* 35–39 (1979).

Weiss, B.F.; Liebschutz, J.L.; Wurtman, R.J.; Munro, H.N.: Participation of dopamine and serotonin receptors in the disaggregation of brain polysomes by *L*-dopa and *L*-5-HTP. J. Neurochem. *24:* 1191–1195 (1975).

Wilk, S.; Mones, R.: Cerebrospinal fluid levels of 3-methoxy-4-hydroxyphenylethylene glycol in parkinsonism before and after treatment with *L*-DOPA. J. Neurochem. *18:* 1771–1773 (1971).

Wirz-Justice, A.: Theoretical and therapeutic potential of indoleamine precursors. Neuropsychobiology *3:* 199–233 (1977).

Young, S.N.; Gauthier, S.; Anderson, G.M.; Purdy, W.C.: Tryptophan, 5-hydroxyindoleacetic acid and indoleacetic acid in human cerebrospinal fluid: interrelationships and the influence of age, sex, epilepsy and anticonvulsant drugs. J. Neurol. Neurosurg. Psychiat. *43:* 438–445 (1980).

Young, S.N.; Sourkes, T.L.: The antidepressant action of tryptophan. Lancet *ii:* 897–898 (1974).

Young, S.N.; Sourkes, T.L.: Tryptophan in the central nervous system: regulation and significance; in Agranoff, Aprison, Advances in neurochemistry, vol. 2, pp. 133–191 (Plenum Press, New York 1977).

S.N. Young, PhD, Department of Psychiatry, McGill University,
1033 Pine Avenue West, Montreal H3A 1A1 (Canada)

Adv. biol. Psychiat., vol. 7, pp. 126–137 (Karger, Basel 1981)

Use of Selective Acetylenic MAO Inhibitors in Depression

M.B.H. Youdim, J.P.M. Finberg
Technion – Israel Institute of Technology, Department of Pharmacology, Faculty of Medicine, Haifa, Israel

Introduction

The MAO inhibitors were among the first group of drugs for which a clear biochemical basis for drug action in the CNS could be demonstrated. The antidepressant properties of the MAO inhibitors were related to an irreversible inhibition of brain MAO, which also caused an increase in the brain levels of noradrenaline (NA), dopamine (DA) and 5-hydroxytryptamine (5-HT) [for review see *Youdim* et al., 1981].

The introduction of new drugs to clinical use tends to follow a well-known cycle. After initial scepticism there is a general and uncritical enthusiasm for the drug's value. Sooner or later this is followed by a recognition that there are somewhat undesirable side effects, and sometimes an equally uncritical reaction against their clinical use. Ultimately, this phase is followed by one of balanced appraisal of the advantages, appropriate uses and dangers of the drug.

The MAO inhibitors have been in clinical use for the past 25 years. The older drugs, iproniazid, phenelzine, isocarboxazid, tranylcypromine and pargyline have passed through all these phases and have entered the final stage where their ultimate place in therapy is becoming clear.

Following the recognition of their side effects, particularly the interaction with sympathomimetic amines, e.g., tyramine, or amine precursors, e.g., *L*-dopa [*Marley,* 1977; *Youdim,* 1977], their clinical use became severely limited. Added to this problem was the report of the Medical Research Council [1965] on a trial of imipramine, electroconvulsive therapy, phenelzine and placebo in depressed patients, which suggested that phenelzine was not more effective than placebo.

However, in the last few years there has been a considerable resurgence of interest in MAO inhibitors, both in use and clinical research. The results of these controlled trials have shown abundant evidence to support the view that MAO inhibitors are effective, and specific antidepressants, when given in adequate doses, for an adequate period of time and to the correct type of patients [*Pare*, 1976; *Nies and Robinson*, 1981; for review see *Youdim and Paykel*, 1981]. Whether the antidepressant action of MAO inhibitors is related to raised brain levels of 5-HT and NA or to down regulation of the β-adrenergic receptor as suggested by *Sulser* [this volume] is a matter of conjecture. What is important is that MAO inhibitors are now unequivocally considered to possess antidepressant activity, but their possible interaction with sympathomimetic amines ('the cheese effect') [*Marley*, 1977] still limits their use as drugs of first choice. The discovery that MAO exists in multiple forms, each with different substrate specificity, inhibitor sensitivity and distribution [*Johnston*, 1968; *Youdim* et al., 1969; *Houslay* et al., 1976], gave impetus to the possibility of developing selective inhibitors without undesirable side effects [*Youdim* et al., 1971]. This paper will deal with biochemical and pharmacological characterization of multiple forms of MAO and their selective inhibitors and the possible use of new selective MAO inhibitor drugs as antidepressants.

Multiple Forms of MAO as an in vivo Reality

The consensus of opinion is that there are two major forms of MAO, type A and type B [*Johnston*, 1968]. Among the substrates of type A one can include 5-HT, adrenaline (A); NA and octopamine. The selective inhibitor of this form is clorgyline [*Johnston*, 1968]. The B form of the enzyme has substrate specificity for benzylamine and phenylethylamine [*Johnston*, 1968; *Yang and Neff*, 1973] and is selectively inhibited by (–)-deprenyl [*Knoll and Magyar*, 1972]. Tyramine and DA are substrates for both forms [*Houslay* et al., 1976]. These criteria have been employed to determine the distribution of the two forms in various tissues of man [*Riederer* et al., 1978] and animals [*Tipton* et al., 1976] and the two selective inhibitors are also used as models for the development of new selective MAO inhibitors. When envisaging the use and development of MAO inhibitors as antidepressants one must consider the distribution of A and B forms of MAO in man. Studies have shown that while 80% of human brain MAO is of the type B [*Squires*, 1972; *Youdim*, 1977; *Riederer* et al., 1978], the A form is the major component in the small intestine [*Squires*, 1972; *Youdim*, 1977], while liver contains equal amounts of both forms. This has led to the concept that the selective B inhibitors could be used to inhibit brain MAO and achieve neuropharmacological effects, while keeping the peripheral MAO-A in the liver and small intestine intact for inactivation of dietary derived tyramine, DA and 5-HT. The distribution of MAO-A and -B in the rat tissues is similar to that in man with the main exception of the brain, where their ratio is 1:1.

It has been reported that the intraneuronal MAO associated with sympathetic nerve endings is of type A [*Neff and Fuentes*, 1976]. Although it has not yet been possible to determine the type of MAO found with the biogenic amine containing neurons of brain, it is possible to demonstrate that different MAOs are responsible for the deamination of various monoamines. After administration of clorgyline to rats, 5-HT, DA and NA increase in the brain, while the administration of deprenyl elevates DA and β-phenylethylamine levels [*Neff* et al., 1974; *Green and Youdim,* 1975; *Green* et al., 1977; *Riederer* et al., 1978, 1981]. Decreased levels of deaminated metabolites are seen in association with the increased brain monoamine concentrations [*Neff* et al., 1974; *Sharman,* 1976; *Maitre* et al., 1976]. The increase in brain amines seen after clorgyline is not as great as that observed after treatment with the non-selective MAO inhibitor, tranylcypromine, or when the two selective inhibitors are combined so that both A and B forms of MAO are fully inhibited [*Green and Youdim,* 1976; *Green* et al., 1977].

When rats are injected with a non-selective MAO inhibitor followed by *L*-dopa [*Everett* et al., 1963] or *L*-tryptophan [*Grahame-Smith,* 1971] a characteristic behavioural syndrome of hyperactivity and stereotypy results. These behavioural syndromes have been related to the increased 'functional activity' of DA or 5-HT resulting from their continued formation but inhibited catabolism within the neuron and their release at post-synaptic receptor sites [*Youdim* et al., 1981]. Using these animal behavioural models, we have examined the biochemical and behavioural action of selective MAO inhibitors. Neither clorgyline nor deprenyl at concentrations which fully inhibited either the A or B form of MAO was able to produce the characteristic behavioural syndrome of hyperactivity observed with the non-selective MAO inhibitor, tranylcypromine, given together with the amine precursors.

When clorgyline and deprenyl were given in combination, at concentrations at which each selectively inhibited by 100% MAO-A or -B, the characteristic syndrome was observed following *L*-dopa or *L*-tryptophan injection.

These results were interpreted in the following manner; (a) In vivo the selectivity of these MAO inhibitors for the A or B enzyme form is not as pronounced as that observed in vitro [*Tipton* et al., 1976]. (b) When only one form of MAO is selectively inhibited in vivo with either clorgyline or (–)-deprenyl, the concentration of amine within the neuron rises satisfying the K_m of the second enzyme which then can deaminate it at a significant rate. This may be one reason why with clorgyline or (–)-deprenyl given separately the concentration of brain DA, NA or 5-HT is not increased to the same extent as is observed following tranylcypromine or administration of the selective inhibitors in combination [*Green and Youdim,* 1975; *Green* et al., 1977; *Squires and Buus Lassen,* 1975]. Recent in vitro studies have confirmed our in vivo findings and shown that either 5-HT or β-phenylethylamine at low concentrations are substrates for their respective MAOs, but at high concentrations they can be deaminated by the other MAO while at intermediate levels they are substrate for both enzymes [*Ekstedt,* 1976; *Achee and Gabay,* 1977; *Kinemuchi* et al., 1979; *Suzuki* et al., 1979; *Kawatra and Sourkes,* 1979; *Tipton and Mantle,* 1981]. (c) An inhibition of at least 85% of both enzyme forms is needed before brain amine levels can rise sufficiently to spill over into 'functional activity'. Thus, the 'functional activity' of DA and 5-HT formed from their precursors may not necessarily be increased in man when using selective inhibitors in a clinical setting. However, *Birkmayer* et al. [1975, 1977] and later others [*Lees* et al., 1977; *Yahr,* 1978, *Rinne* et al., 1978] showed that deprenyl, the selective MAO-B inhibitor, is beneficial in Parkinson's disease when used as a supplement to *L*-dopa therapy, especially in patients displaying 'on-off' phenomena and in those subjects who had lost their response to *L*-dopa. The anti-parkinsonian action of deprenyl was suggested to be related to the inhibition of the major component of brain enzyme [*Birkmayer* et al., 1977], namely MAO-B for which DA is a selective substrate in human brain [*Glover* et al., 1977], thus making more DA available for neurotransmission. The proof of this hypothesis has come from brain autopsy studies [*Riederer* et al., 1978]. The daily dose of deprenyl (10 mg) has been sufficient to inhibit MAO-B selectively and almost completely in 17 specific brain areas and raise brain levels of DA, in comparison with brains from patients treated with *L*-dopa only. It is worth noting that in an earlier study where brains were obtained from geriatric patients with terminal diseases, but were treated with MAO inhibitors (clorgyline, tranylcypromine or isocarboxazid) for their depression, only clorgyline treatment failed to induce significant human brain MAO inhibition [*Youdim* et al., 1972]. These results further support the view that the major component of human brain MAO is type B [*Riederer* et al., 1978].

Type A versus Type B Inhibitors as Antidepressants

The selective MAO type A and type B inhibitors are not in clinical use as antidepressants. However, a number of controlled and open studies have been carried out with both classes of drugs [*Murphy* et al., 1979; *Mendlewicz and Youdim,* 1978; *Youdim* et al., 1981; *Youdim and Paykel,* 1981; *Mann and Gershon,* 1980]. The picture is far from clear as to which type of selective inhibitor is an effective antidepressant, since positive results have been obtained with both classes of drugs. *Murphy* et al. [1979] suggest that MAO type A inhibitors (clorgyline and Lilly 51641) are more suitable, since type A enzyme is located intraneuronally and its substrates include 5-HT and NA. Selective type A inhibitors are also very effective in reversing the pharmacological and depressant actions of reserpine and tetrabenazine, while pargyline and deprenyl (the selective B inhibitors) are not.

The major drawback in the use of MAO inhibitors as antidepressants is their interaction with sympathomimetic amines [*Marley,* 1977] resulting in hypertensive crises ('cheese effect'). This limiting side effect is also associated with clorgyline but not with deprenyl [*Knoll,* 1976]. The fact that deprenyl does not potentiate the sympathomimetic action of orally or intravenously administered tyramine [*Elsworth* et al., 1978] or *L*-dopa in man together with the report that the major component of human brain MAO is type B, should make drugs like deprenyl effective and safe antidepressants. Both selective MAO-B inhibitors, deprenyl and pargyline, have been reported to have antidepressant activity [*Varga and Tringer,* 1967; *Kline,* 1963; *Mann and Gershon,* 1980]. Deprenyl is also capable of potentiating the antidepressant activity of 5-hydroxytryptophan (5-HTP) [*Mendlewicz and Youdim,* 1980]. The antidepressant activity of 5-HTP is, in all probability, related to the formation of 5-HT. The latter's accumulation within the neuron and at the synaptic cleft is due to inhibition of MAO [*van Praag,* 1979]. How can one explain the contradictory action of an MAO-B inhibitor to potentiate the 'functional activity' and antidepressant action of the MAO-A substrate, 5-HT? It should be remembered that the selectivity of MAO-A or -B towards their respective substrates is not absolute. This is best explained by the results that highly purified MAO-B from beef liver and human platelets is capable of deaminating a substantial amount of MAO-A substrates such as 5-HT and NA [*Salach* et al., 1979; *Youdim and Hefez,* 1981]. Furthermore, in enzyme preparations where both A and B activity exist, as in the case of human brain, a substantial portion of 5-HT deamination is the result of MAO-B activity [*Tipton and Mantle,* 1981]. Considering that 80% of human brain MAO activity is of type B one can envisage why deprenyl should have antidepressant activity. It should also be remembered that if given in higher doses the selectivity of these inhibitors is lost [*Waldmeier* et al., 1981]. This may very well have happened in the trial studies of *Mendlewicz and Youdim* [1980] and *Mann and Gershon* [1980], since these authors used a regimen of 15 mg (–)-deprenyl daily, which is higher than that employed as an adjunct to *L*-dopa treatment of akinesia in parkinsonian subjects.

Selective MAO-A or -B Inhibitors and the 'Cheese Effect'.

The inhibitors of MAO having an N-propargyl (acetylenic) moiety offer the best conditions to design selective drugs without the 'cheese effect'. These compounds bind irreversibly with N-5 of FAD at the active site [*Salach* et al., 1979]. Structure-activity studies with aromatic N-propargyl derivatives have shown that since these compounds resemble the enzyme substrates (fig. 1) the distance between the aromatic ring and the N-propargyl terminal is most crucial in designating type A or type B MAO inhibitory property. For MAO type B selective inhibition, a distance equivalent to 1 or 2 carbon units is required [*Knoll* et al., 1978; *Kalir* et al., 1981]. The suggestion by *Knoll* et al. [1978] that the active site of type A MAO is best suited for the attachment of indole component has been re-

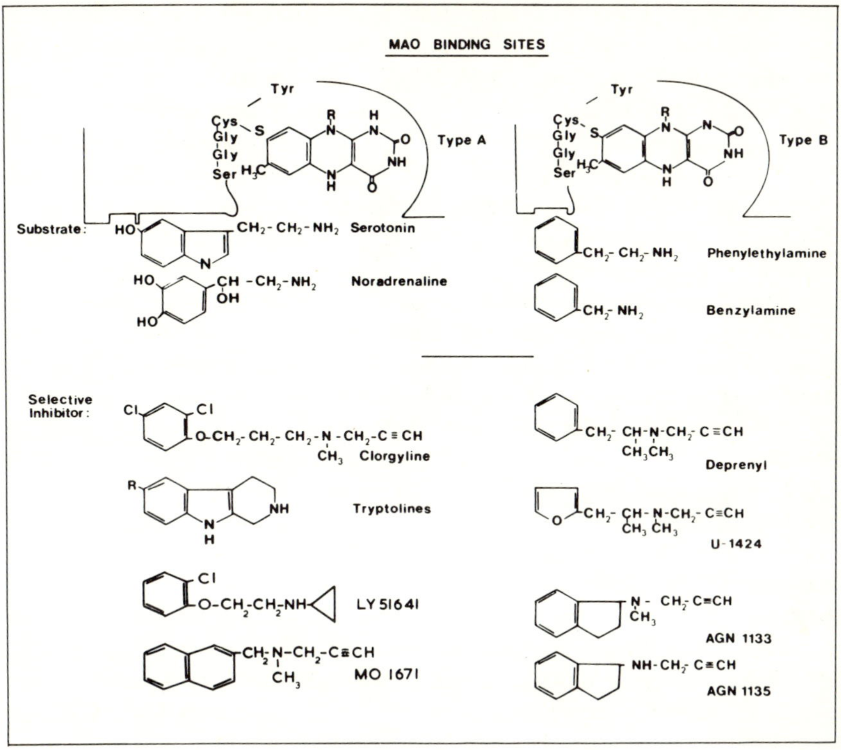

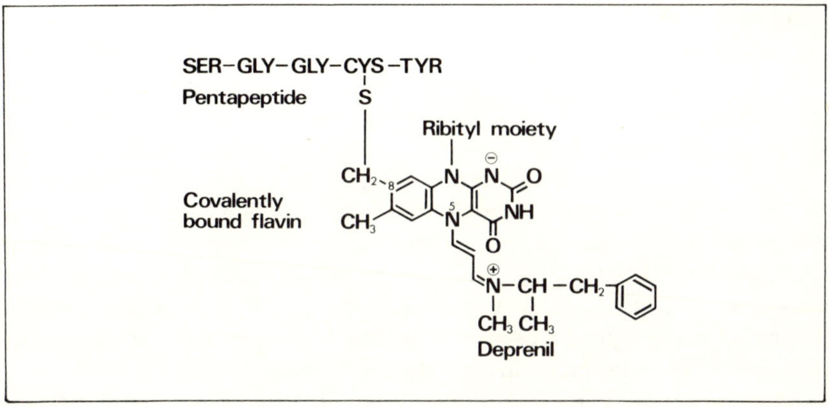

Fig. 1. The schematic structure of the active sites of MAO type A and type B, showing the covalent attachment of the enzyme flavin co-factor. The acetylenic inhibitors bind irreversibly to the N-5 of the flavin [*Salach* et al., 1979].

futed by *Kalir* et al. [1981]. The latter authors have shown that an indole moiety is not a prerequisite for selective MAO type A inhibition [although indole amine derivatives are selective MAO type A inhibitors; *Youdim and Oppenheim,* 1981], and have suggested that the distance between the N-propargyl and aromatic moiety needs to be equivalent to at least 3 or 4 carbon units as in clorgyline, M & B 9303 and MO 1671 (fig. 1). Therefore, the smaller active site of MAO type B may be more restricted. This may be one explanation as to why the selectivity of MAO-A inhibitors for the A enzyme is more pronounced than that of MAO-B inhibitors for the B enzyme.

A number of reasons have been proposed to explain the potentiation of indirectly acting sympathomimetic amines such as tyramine by MAO inhibitors. These include (a) inhibition of sympathomimetic amine metabolism in the gastrointestinal tract and liver [*Blackwell and Marley,* 1966; *Horwitz* et al., 1964; *Marley,* 1977]; (b) inhibition of intraneuronal sympathomimetic amine metabolism [*Goldberg,* 1962; *Trendelenburg,* 1972], and (c) inhibition of metabolism of NA released by the indirectly acting amine [*Trendelenburg,* 1972].

Evidence that inhibition of extraneuronal metabolism can lead to potentiation of an indirectly acting amine was provided in the experiments of *Knoll* [1976] with the cat nictitating membrane. The contractile effect of β-phenylethylamine on this organ was greatly potentiated by exclusion of the liver from the circulation. Blood levels of tyramine were also shown to be elevated following MAO inhibition in the dog [*Faraj* et al., 1977]. The fact that the cardiovascular effects of indirectly acting amines which are not substrates for MAO (e.g., amphetamine) are potentiated by a selective type A but not by a type B inhibitor [*Simpson,* 1978a] indicates that a reduced breakdown of the indirectly acting amine itself is not the only reason for the 'cheese effect', but prolongation of the action of released NA is also an important factor. This evidence also suggests that any selective type A inhibitor may be expected to elicit a cheese reaction, since the type A enzyme is the form associated with sympathetic nerves (see above). Further evidence for the importance of inhibition of neuronal MAO in production of the 'cheese effect' is provided by the experiments of *Sandler* et al. [1980] in the pig. In this animal, tissue MAO is largely type B, and (–)-deprenyl is considerably more effective than clorgyline in inhibition of tyramine oxidation. Neuronal MAO in this species, however, as well as in other species investigated, may be largely type A. The fact that the pressor effect of tyramine in the pig was potentiated by clorgyline but not by (–)-deprenyl shows that inhibition of neuronal MAO, and not necessarily reduction in extraneuronal tyramine metabolism, is the important factor in eliciting this response.

These concepts are borne out by the experiments of *Finberg* et al. [1981a] using the isolated rat vas deferens. Concentrations of (–)-deprenyl and AGN 1135 which selectively inhibit MAO type B did not potentiate the effects of tyramine on the vas deferens, whereas concentrations of clorgyline selectively in-

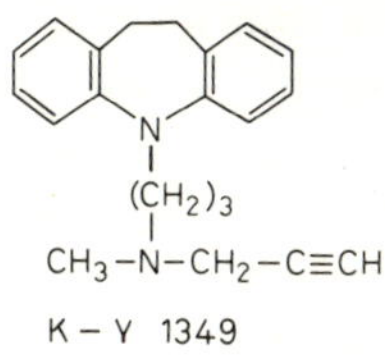

Fig. 2. The structure of the selective MAO type A inhibitor (K-Y 1349; N-propargylimipramine) [*Kalir* et al., 1981].

hibiting MAO type A did. When higher concentrations of (–)-deprenyl and AGN 1135, which inhibit both MAO types A and B were used, potentiation of the tyramine-induced contractions could be demonstrated; however, at these high drug concentrations, an additional reversible effect to inhibit the tyramine response is present, and potentiation of tyramine was seen only after the inhibitors were washed out of the organ bath.

Potentiation of the β-phenylethylamine effect on the nictitating membrane by (–)-deprenyl [*Knoll,* 1976] appears to be mainly due to inhibition of metabolism of the indirectly acting amine itself, since clorgyline causes a much weaker potentiation of this effect. The lack of potentiation of the pressor effect of β-phenylethylamine by (–)-deprenyl is as yet unexplained, but may be related to the rapidly developing tachyphylaxis to the pressor response elicited by this amine. In a setting more closely related to the actual clinical situation, i.e., potentiation of the pressor effect of orally administered tyramine by clorgyline [*Lader* et al., 1970], all of the above-described mechanisms may be involved.

Non-potentiation of the effect of orally administered tyramine by (–)-deprenyl has been related to both lack of inhibition of gastro-intestinal tract MAO, and lack of inhibition of neuronal MAO [*Knoll,* 1976; *Finberg* et al., 1981a, b]. A third mechanism, inhibition of neuronal uptake of tyramine, has been invoked by *Knoll* [1978]. (–)-Deprenyl, however, is a relatively weak inhibitor of amine uptake in vas deferens [*Finberg* et al., 1981a, b] and heart [*Simpson,* 1978b], and concentrations higher than those required for complete inhibition of MAO-B are needed for effective reduction of amine uptake. Neuronal elements associated with blood vessels may be more sensitive to the uptake-inhibitory action of (–)-deprenyl, or, alternatively, inhibition of MAO-B may not be associated with tyramine potentiation. Further experiments are required to prove this hypothesis.

At the time of writing, no efficient irreversible inhibitor of MAO-A or A and B has been described which is free of the 'cheese effect'. One way in which such a drug could be developed is by incorporation of MAO inhibitory and amine uptake inhibitory properties in the same molecule.

An example of this approach is shown by the experimental drug K 1349 (fig. 2), which we have synthesized. This compound, which represents a synthesis of desmethylimipramine with the propargylamine enzyme-inactivating group, is a selective MAO-A inhibitor with an IC_{50} of 10^{-7} *M* and 10^{-3} *M* (using ^{14}C-5-HT and phenylethylamine as substrates, respectively, and rat brain as enzyme source). This compound also inhibited tyramine responses in the isolated rat vas deferens (10^{-7} *M*), although noradrenaline and angiotensin responses were also inhibited at a slightly higher concentration. The detailed pharmacology of K 1349 remains to be determined, but it represents an experimental approach to the design of antidepressant MAO-A inhibitors without the 'cheese effect'.

References

Achee, F.M.; Gabay, S.: Studies on monoamine oxidase: inhibition of bovine brain MAO in intact mitochondria by selective inhibitors. Biochem. Pharmac. *26:* 1637–1644 (1977).

Birkmayer, W.; Riederer, P.; Ambrozi, L.; Youdim, M.B.H.: Implications of combined treatment with 'Madopar' and *L*-deprenyl in Parkinson's disease. A long-term therapy. Lancet *i:* 439–443 (1977).

Birkmayer, W.; Riederer, P.; Youdim, M.B.H.; Linauer, W.: Potentiation of anti-akinetic effect after *L*-dopa treatment by an inhibitor of MAO-B-deprenyl. J. neural Transm. *36:* 303–323 (1975).

Blackwell, B.; Marley, E.: Interactions of cheese and of its constituents with monoamine oxidase inhibitors. Br. J. Pharmac. Chemother. *26:* 120–141 (1966).

Ekstedt, G.: Substrate specificity of the different forms of monoamine oxidase in rat liver mitochondria. Biochem. Pharmac. *25:* 1133–1138 (1976).

Elsworth, J.D.; Glover, V.; Reynolds, G.P.; Sandler, M.; Lees, A.J.; Phuapradit, P.; Shaw, K.M.; Stern, G.M.; Kumar, P.: Deprenyl administration in man. A selective MAO-B inhibitor without the 'cheese effect'. Psychopharmacology, Berlin *57:* 33–38 (1978).

Everett, G.M.; Wiegand, R.G.; Rinaldi, F.U.: Pharmacologic studies of some nonhydrazine MAO inhibitors. Ann. N.Y. Acad. Sci. *107:* 1068–1080 (1963).

Faraj, B.A.; Dayton, P.G.; Camp, V.M.; Wilson, J.P.; Malveaux, E.J.; Schlant, R.C.: Studies on the fate of tyramine in dogs: the effect of monoamine oxidase inhibition, portafemoral shunt and coronary artery ligation on the kinetics of tyramine. J. Pharmac. exp. Ther. *200:* 384–393 (1977).

Finberg, J.P.M.; Tenne, M.; Youdim, M.B.H.: Tyramine antagonistic properties of AGN 1135, an irreversible inhibitor of monoamine oxidase, type B. Br. J. Pharmacol. (in press 1981a).

Finberg, J.P.M.; Tenne, M.; Youdim, M.B.H.: Selective irreversible propargyl derivative inhibitors of monoamine oxidase (MAO) without the cheese effect; in Youdim, Paykel, Monoamine oxidase inhibitors: the state of art, pp. 31–43 (Wiley, Chichester 1981b).

Glover, V.; Sandler, M.; Owen, F.; Riley, G.J.: Dopamine is a monoamine oxidase B substrate in man. Nature *265;* 80–81 (1977).

Goldberg, L.I.: Monoamine oxidase inhibitors. J. Am. med. Ass. *190:* 456–462 (1962).

Grahame-Smith, D.G.: Studies in vitro on the relationship between brain tryptophan, brain 5-HT synthesis and hyperactivity in rats treated with a monoamine oxidase inhibitor and *L*-tryptophan. J. Neurochem. *18:* 1053–1066 (1971).

Green, A.R.; Mitchell, B.D.; Tordoff, A.F.C.; Youdim, M.B.H.: The evidence for dopamine deamination by both type A and type B monoamine oxidase in rat brain in vivo and for degree of inhibition of enzyme necessary for increased functional activity of dopamine and 5-hydroxytryptamine. Br. J. Pharmacol. *60:* 343–349 (1977).

Green, A.R.; Youdim, M.B.H.: Effects of monoamine oxidase inhibition by clorgyline, deprenyl and tranylcypromine on 5-hydroxytryptamine concentration in rat brain and hyperactivity following tryptophan administration. Br. J. Pharmacol. *55:* 415–422 (1975).

Green, A.R.; Youdim, M.B.H.: Use of behavioural model to study the action of monoamine oxidase inhibition in vivo; in Wolstenholme, Knight, Ciba Fd. Symp. No. 39, pp. 231–247 (Elsevier, Amsterdam 1976).

Horwitz, D.; Lovenberg, W.; Engelman, K.; Sjverdsma, A.: Monoamine oxidase inhibitors, tyramine and cheese. J. Am. med. Ass. *188:* 1108–1100 (1964).

Houslay, M.D.; Tipton, K.F.; Youdim, M.B.H.: Multiple forms of monoamine oxidase: fact and artefact. Life Sci. *19:* 467–473 (1976).

Johnston, J.P.: Some observations upon a new inhibitor of monoamine oxidase in brain tissue. Biochem. Pharmac. *17:* 1285–1297 (1968).

Kalir, A.; Sabbagh, A.; Youdim, M.B.H.: Structure activity relationships of selective acetylenic suicide inhibitors of MAO type B. Br. J. Pharmacol. (in press 1981).

Kawatra, M.M.; Sourkes, T.L.: Muscle and liver monoamine oxidases: effects of iron and riboflavin deficiencies in the rat; in Singer, Von Korff, Murphy, Monoamine oxidase structure, function and altered functions, pp. 279–284 (Academic Press, New York 1979).

Kinemuchi, H.; Wakui, Y.; Toyoshima, Y.; Hayashi, N.; Kamijo, K.: β-Phenylethylamine (PEA), a concentration-dependent preferential substrate for multiple forms of MAO; in Singer, Von Korff, Murphy, Monoamine oxidase, structure, function and altered functions, pp. 205–212 (Academic Press, New York 1979).

Kline, N.S.: Use of pargyline (Eutonyl) in private practice. Ann. N.Y. Acad. Sci. *107:* 1090–1106 (1963).

Knoll, J.: Analysis of the pharmacological effects of selective monoamine oxidase inhibitors; in Wolstenholme, Knight. Ciba Symp., No. 39, pp. 135–155 (Elsevier, Amsterdam 1976).

Knoll, J.: The possible mechanisms of action of (–)-deprenyl in Parkinson's disease. J. neural transm. *43:* 177–198 (1978).

Knoll, J.; Ecsery, Z.; Magyar, K.; Satory, E.: Novel (–)-deprenyl-derived selective inhibitors of B-type monoamine oxidase. The relation of structure to their action. Biochem. Pharmac. *27:* 1739–1747 (1978).

Knoll, J.; Magyar, K.: Some puzzling pharmacological effects of monoamine oxidase inhibitors. Adv. biochem. Psychopharmacol. *5:* 393–408 (1972).

Lader, M.H.; Sakalis, G.; Tansella, M.: Interactions between sympathomimetic amines and a new monoamine oxidase inhibitor. Psychopharmacology, Berlin *18:* 118–123 (1970).

Lees, A.J.; Shaw, K.M.; Kohout, L.J.; Stern, G.M.; Elsworth, J.D.; Sandler, M.; Youdim, M.B.H.: Deprenil in Parkinson's disease. Lancet *ii:* 791–795 (1977).

Maitre, L.; Delini-Stula, A.; Waldmeier, P.C.: Relations between the degree of monoamine oxidase inhibition and some psychopharmacological responses to monoamine oxidase inhibitors in rats; in Wolstenholme, Knight, Ciba Fdn Symp., No. 39, pp. 247–271 (Elsevier, Amsterdam 1976).

Mann, A.; Gershon, S.: *L*-Deprenyl, a selective monoamine oxidase type B inhibitor in endogenous depression. Life Sci. *26:* 877–882 (1980).

Marley, E.: Monoamine oxidase inhibitors and drug interactions; in Grahame-Smith, Drug interactions, pp. 171–194 (MacMillan, London 1977).

Medical Research Council: Clinical trial of the treatment of depressive illness. Br. med. J. *i:* 881–886 (1965).

Mendlewicz, J.; Youdim, M.B.H.: Anti-depressant potentiation of 5-hydroxytryptophan by *L*-deprenyl, an MAO 'type B' inhibitor. J. neural Transm. *43:* 279–286 (1978).

Mendlewicz, J.; Youdim, M.B.H.: Anti-depressant potentiation of 5-hydroxytryptophan by *L*-deprenyl in affective illness. J. aff. Disord. *2:* 137–146 (1980).

Murphy, D.L.; Lipper, S.; Campbell, I.C.; Major, L.F.; Slater, S.; Buchsbaum, M.S.: Comparative studies of MAO-A and MAO-B inhibitors in man; in Singer, Von Korff, Murphy, Monoamine oxidase: structure, function and altered function, pp. 457–576 (Academic Press, New York 1979).

Neff, N.H.; Fuentes, A.: The use of selective monoamine oxidase inhibitor drugs for evaluating pharmacological and physiological mechanisms; in Wolstenholme, Knight, Monoamine oxidase and its inhibition. Ciba Fdn Symp., No. 39, pp. 163–173 (Elsevier, Amsterdam 1976).

Neff, N.H.; Yang, H-Y.T.; Fuentes, J.A.: The use of selective monoamine oxidase inhibitor drugs to modify amine metabolism in brain. Adv. biochem. Psychopharmacol. *12:* 49–58 (1974).

Nies, A.; Robinson, D.: Comparison of clinical effects of amitriptyline and phenelzine treatment; in Youdim, Paykel, Monoamine oxidase inhibitors: the state of the art, pp. 141–148 (Wiley, Chichester 1981).

Pare, C.M.B.: Introduction to clinical aspects of monoamine inhibitors in the treatment of depression; in Wolstenholme, Knight, Monoamine oxidase and its inhibition. Ciba Fdn Symp., No. 39, pp. 271–296 (Elsevier, Amsterdam 1976).

Praag, H.M. van: Serotonin and pathogenesis of affective disorders; in Youdim, Aromatic amino acid hydroxylases and mental disease, pp. 299–378 (Wiley, Chichester 1979).

Riederer, P.; Reynolds, G.; Youdim, M.B.H.: Selectivity of MAO inhibitors in human brain and their clinical consequences; in Monoamine oxidase inhibitors: the state of the art, pp. 63–76 (Wiley, Chichester 1981).

Riederer, P.; Youdim, M.B.H.; Rausch, W.D.; Birkmayer, W.; Jellinger, K.; Seemann, D.: On the mode of action of *L*-deprenyl in the human central nervous system. J. neural transm. *43:* 217–226 (1978).

Rinne, U.K.; Sirrtola, T.; Sonninven, V.: (–)-Deprenyl treatment of 'on-off' phenomena in Parkinson's disease. J. neural Transm. *43:* 253–262 (1978).

Salach, J.I.; Detmer, K.; Youdim, M.B.H.: The reaction of bovine and rat liver monoamine oxidase with (^{14}C)-deprenyl. Molec. Pharmacol. *16:* 234–241 (1979).

Sandler, M.; Glover, V.; Ashford, A.; Esmail, A.: The inhibition of tyramine oxidation and the tyramine hypertensive response ('cheese effect') may by independent phenomena. J. neural Transm. *48:* 241–247 (1980).

Sharman, D.F.: Can the intra and extra-homoneuronal metabolism of catecholamines be distinguished in the mammalian central nervous system; in Wolstenholme, Knight, Monoamine oxidase and its inhibition. Ciba Fdn Symp. No. 39, pp. 203–229 (Elsevier, Amsterdam 1976).

Simpson, L.L.: Mechanism of the adverse interaction between monoamine oxidase inhibitors and amphetamine. J. Pharmac. exp. Ther. *205:* 392–399 (1978a).

Simpson, L.L.: Evidence that deprenyl, a type B monoamine oxidase inhibitor, is an indirectly acting sympathomimetic amine. Biochem. Pharmac. *27:* 1591–1595 (1978b).

Squires, R.F.: Multiple forms of monoamine oxidase in intact mitochondria as characterized by selective inhibitors and thermal stability: a comparison of eight mammalian species. Adv. biochem. Psychopharmacol. *5:* 355–370 (1972).

Squires, R.F.; Buus Lassen, J.C.: The inhibition of A and B forms of MAO in the production of a characteristic behavioural syndrome in rats after *L*-tryptophan loading. Psychopharmacologia *41:* 145–151 (1975).

Suzuki, O.; Katsumata, Y.; Oya, M.: Characterization of some biogenic monoamines as substrates for type A and type B monoamine oxidase; in Singer, Von Korff, Murphy, Monoamine oxidase: structure, function and altered function, pp. 197–204 (Academic Press, New York 1979).

Tipton, K.F.; Houslay, M.D.; Mantle, T.J.: The nature and locations of the multiple forms of monoamine oxidase; in Wolstenholme, Knight, Monoamine oxidase and its inhibition. Ciba Fdn Symp., No. 39, pp. 5–31 (Elsevier, Amsterdam 1976).

Tipton, K.F.; Mantle, T.J.: Inhibition of rat liver monoamine oxidase by clorgyline and deprenyl; in Youdim, Paykel, Monoamine oxidase inhibitors: the state of the art, pp. 3–15 (Wiley, Chichester 1981).

Trendelenburg, U.: Factors influencing the concentration of catecholamines at the receptors; in Muscholl, Blaschko, Catecholamines, pp. 726–761 (Springer, Berlin 1972).

Varga, E.; Tringer, L.: Clinical trial of a new type promptly acting psychoenergetic agent (phenyl-isopropylmethyl-propynylamine-HC1, E-250). Acta. med. hung. *23:* 289–295 (1967).

Waldmeier, P.C.; Felner, A.E.; Maitre, L.: Long-term effects on selective MAO inhibitors on MAO activity and amine metabolism; in Youdim, Paykel, Monoamine oxidase inhibitors: the state of the art, pp. 87–102 (Wiley, Chichester 1981).

Yahr, M.D.: Overview of present day treatment of Parkinson's disease. J. neural Transm. *43:* 227–238 (1978).

Yang, H.Y.; Neff, N.H.: β-Phenylethylamine: a specific substrate of type B monoamine oxidase of brain. J. Pharmacol. exp. Ther. *187:* 365–371 (1973).

Youdim, M.B.H.: Tyramine and psychiatric disorders; in Neuroregulators and psychiatric disorders, pp. 57–67 (Oxford University Press, New York 1977).

Youdim, M.B.H.; Collins, G.G.S.; Sandler, M.: Multiple forms of rat brain monoamine oxidase. Nature *223:* 626–628 (1969).

Youdim, M.B.H.; Collins, G.G.S.; Sandler, M.: Monoamine oxidase, multiple forms and selective inhibitors. Biochem. J. *121:* 34–36 (1971).

Youdim, M.B.H.; Collins, G.G.S.; Sandler, M.; Jones Bevan, A.B.; Pare, C.M.B.; Nicholson, W.J.: Human brain monoamine oxidase: multiple forms and selective inhibitors. Nature *236:* 225–228 (1972).

Youdim, M.B.H.; Hefez, A.: Platelet function and MAO activity in psychiatric disorders; in Rotman, Platelets: cellular response mechanisms and their biological significance, pp. 289–306 (Wiley, Chichester 1981).

Youdim, M.B.H.; Oppenheim, B.: The effect of 1,2,3,4-tetra-hydro-β-carbolines on monoamine metabolism in the human platelet and platelet aggregation. Neuroscience (in press 1981).

Youdim, M.B.H.; Paykel, E.S.: Monoamine oxidase inhibitors: the state of the art (Wiley, Chichester 1981).

Youdim, M.B.H.; Tipton, K.F.; Finberg, J.P.M.: Monoamine oxidase; in Trendelenburg, Weiner, Handbook of experimental pharmacology: catecholamines II. (Springer, Berlin, in press 1981).

M.B.H. Youdim, MD, Technion – Israel Institute of Technology,
Department of Pharmacology, Faculty of Medicine, Haifa (Israel)

Adv. biol. Psychiat., vol. 7, pp. 138–153 (Karger, Basel 1981)

Sulpiride, Psychosis, and Mood

P. Borenstein, F. Gekiere
Hôpitaux Psychiatriques de la Seine, Paris, France

Introduction

During recent symposia on the neuroleptics emphasis was placed on the originality of the benzamides and the need for precise knowledge of their chemical, experimental, and psychopharmacological mechanisms of action in relation to their clinical use.

Though it is evident that the three principal molecules available for clinical administration at the present time, sulpiride, sultopride, and tiapride, as well as the molecules still being developed, constitute a new family of psychotropic agents, it is still very debatable as to whether they can be included in the class of neuroleptics for which they possess only one or two pharmacological properties. The most controversial question is the possible antidepressant activity of sulpiride, a compound with dual activity, though it is widely accepted that the benzamides possess unquestionable disinhibitory effects apart from their sedative and antipsychotic actions.

Does this effect on mood (which often intervenesinits action on anxiety) arise from stimulating activity, and perhaps only then by an action on alertness? It is related only to its antidelirium properties (antipsychotic action which is also sometimes questioned)? These are the questions we will try to elucidate.

After a rapid review of the preliminary studies conducted [*Borenstein,* 1967; *Borenstein* et al., 1968a, b; *Borenstein and Gekiere,* 1979] we shall discuss the results of the clinical trials carried out over the last 12 years in an attempt to extract the positive and negative arguments concerning the antidepressant effects of sulpiride. We shall also try to clarify the terms such as depression, antidepressant, inhibition, disinhibitory, anxiety, anxiolytic, sedative – terms which often lead to confusion because of their ambiguity. After this review of the principal French and foreign publications (summarized as much as possible) we shall present several typical case reports which illustrate the action of sulpiride on the target symptom of mood in patients with various psychotic disorders. Side-effects and contraindications will also be discussed.

Finally, the discussion will be centered on recent nosographical data relating to depressive states, their therapy, and the classification of psychotropic agents, taking into account the most up-to-date information available in the fields of pharmacology, neurobiochemistry, endocrinology, and neurophysiology.

Review of French and Foreign Publications

This is a difficult task as very little is left to be said after 15 years of use of this product which has been the subject of many debates, sometimes, regrettably rather too passionate. It would be easy to define the product by two key words 'antipsychotic – thymoanaleptic' as reported in many texts, but one thesis is not sufficient to analyze and prepare a synthesis of all the clinical studies conducted on sulpiride, from the time of its initial indications in gastroenterology [*Cornet and Grivaux,* 1968, 1969], including its endocrine impact, while in reviewing the whole range of neuropsychiatric indications. After initial screening lasting 2 years, in 1967 we prepared a clinical evaluation of the product which possessed the original factors differentiating it from conventional psychotropic agents by its chemical and pharmacological properties. These studies, in 1968 and 1969, defined and clarified the indications and contraindications. (320 patients were administered doses of 150–1,200 or even 1,500 mg/day either by intramuscular injection or orally, doses which appeared to be immediately effective, for periods ranging from 1 month to 2 years.) At this stage, we were already impressed by the dual nature of the effect on those target symptoms which reacted best to the compound: on the psychotic process on the one hand and on the emotional, affective, and mood spheres on the other. We emphasized its rapid action, within the first few days, on mood and anxiety and stated, and I quote: 'The anxiolytic effect is related in time with the thymoanaleptic effect, in the absence of which it was never observed. They both occur at an early stage, and both effects precede the action on delirious activity in chronic delirium patients.'

In fact, this favourable effect on hallucinatory symptoms and delusional themes, when we observe them, was noticeable early in the 2nd week of treatment. Those syndromes obtaining the maximum benefit from this therapy appeared to us to be: simple schizophrenias, paranoid schizophrenias with major depressive symptoms, chronic persecutory delusional states, and all types of depression, with the reservation, however, that severe depression requires adequate doses of amitriptyline, while the dosages of both products could be reduced. Finally, the infrequency of side-effects, especially of the extrapyramidal type, was a further reason for us to prescribe sulpiride whenever delirium and/or depression was present, to allow us to prescribe, as maintenance therapy to ambulatory patients, a single product which never or only infrequently required antiparkinsonian agents and only rarely caused disturbances in alertness. It is interesting to note that during this period we observed endocrine-type secondary phenomena (amenorrhoea, galactorrhoea), but to a much less obvious degree than during the following years. Before discussing this in greater detail we would like to present the principal data concerning the activity of sulpiride reported in the published literature over the last 12 years. In view of the very large number of reports published on sulpiride both in France and abroad, we shall discuss only a selection of those articles relevant to its mode of action, through its indications and contraindications.

From the French publications which mention the activity of sulpiride on the psychoses and mood disturbances we have obtained two principal notions. Firstly, rating scales and objective test are rarely employed, and secondly the majority of the studies are open ones and involve sulpiride treatment only. When psychometric tests were conducted, they were usually based on the BPRS. *Carrere* [1968] reported 75% of positive results in approximately 40 psychotic patients, with efficacy in the following symptoms (in decreasing order): anxiety, tension, depressive tendencies, hostility, hallucinatory behaviour, lack of cooperation, peculiar thoughts. He concludes: 'Sulpiride has been shown to be an incisive neuroleptic with associated euphoric effects.' Other authors, however, such as *Volmat* et al. [1969] and *Ropert* [1977], emphasize the stimulating and disinhibitory actions, and discount any thymoanaleptic activity in the proper sense of the term, arguing that there is no inversion of mood. We shall discuss this important point further on. It should be noted, however, at this stage, that *Volmat* et al. [1969] reported overall results in cases in which associations with neuroleptics or thymoanaleptics were employed and that the studies by *Ropert* et al. [1970a, b] were conducted with high doses (600–800 mg/day i.m., 1,200–1,400, and even up to 2,400 mg/day orally). Their conclusions, therefore, are either very positive and demonstrate definite antidepressant and antipsychotic effects or much more reserved and stress the disinhibitory activity only.

As far as foreign publications are concerned, it is interesting to note that after having been the object of numerous studies in most European countries, Africa (Pretoria), and particularly Japan, sulpiride has been studied in only three regions (Canada, Louisiana, USA, and Mexico) in the American continent. Most studies were conducted under controlled conditions using objective rating scales (BPRS, IMPS, NOSIE, Hamilton, and Wittenborn).

Similar divergences of opinion are reported from our Belgian neighbours [*Collard,* 1969, 1976; *Collard* et al., 1971; *De Buck,* 1979a, b]. Italian articles are categoric as to a very clear action on mood, inhibition, and psychosis [*Cicchetti* et al., 1972; *Cassano* et al., 1975]. Studies in Spain, Poland, Israel, Finland, and England are mainly double-blind, controlled comparative trials against amitriptyline, and the results were also positive [*Niskanen* et al., 1975; *Aylward* et al., unpublished] with greater improvement against psychomotor slowing under sulpiride, and fewer side-effects. The Japanese double-blind comparative trials are greatest in number. Objectively controlled by the use of rating scales they also mention a definite antidepressant effect [*Yura and Kato,* 1976: 'superior results with sulpiride during the 1st and 2nd week on depressed mood, while imipramine has a better effect on insomnia, but has more side-effects'.] In actual fact, hundreds of studies have been conducted throughout the world and thousands of patients have been treated for a wide variety of symptoms and at varying dosages. Whether by intramuscular or oral administration, in open or controlled studies, single or double-blind, against placebo or psychotropic agents (neuroleptics or thymoanaleptics), studies were conducted using dosages from 100 to 1,500 mg daily in most cases, but increasing to even 3,600 mg/day [*Collard* et al., 1971], with an average of 150–200 mg/day for low doses and 800–1,000 mg/day for the higher doses. Patients of both sexes have been treated, at ages varying from 20 to 60 years generally, but both elderly patients over 60 years of age [*Josserand,* 1978] as well as infants and adolescents [*Abe,* 1975, 1979] have received therapy. It is difficult to analyze the different indications for which sulpiride was prescribed because of the nosological differences encountered between the different countries, and target symptoms will, therefore, be the basis of our analysis. Associated therapy was administered to variable degrees: sometimes systematically and as initial treatment making it difficult to assess the therapeutic effect of sulpiride alone, especially when a conventional antidepressant, usually amitriptyline, was associated. (This invites comments on the induction of therapeutic attitudes following data obtained during preliminary studies.) The association of another neuroleptic, either initially or at a later stage, also interferes with the evaluation of specific antipsychotic activity. In all cases where associated therapy was administered, the dose of sulpiride and that of the associated product was greatly reduced. Less 'disturbing' is the systematic association of hypnotics or transquillizers (benzodiazepine type), though in this case it is difficult to assess the effect of sulpiride on alertness and anxiety. The same applies to the indications for antiparkinsonian treatment, which few practitioners prescribe only after the appearance of neurological disorders.

Taking these various methodological considerations into account, what are the main criteria for assessing activity among this heterogeneous collection of data? Though some few authors reject any antidepressant and/or psycholeptic activity for sulpiride, most French and foreign clinicians agree that sulpiride possesses favourable effects on the target symptoms of mood, inhibition, anxiety, and psychotic syndromes:

Mood

Whatever the dose, but particularly those below 400 mg, there is a positive effect on moral suffering, sadness, morosity, disgust with life, self-depreciation, self-punishment, ideas of incurability and of suicide, the patient developing a feeling of well-being, euphoria, and relaxation after 1 week of treatment. The product is then qualified as being thymoanaleptic, thymeretic, euthymic, or more rarely thymoregulatory.

Inhibition

It is very difficult to assess the action on psychic inhibition and psychomotor slowing, and the effect on mood and anxiety, as these symptoms are often intricated. A patient may be anxious and not depressed, however, or anxious but not inhibited. Furthermore, a disinhibitory effect is not always associated with anxiolysis and/or euphoria, as with alcohol at small doses for example. On the contrary, disinhibition without anxiolysis and with persistence of sadness is frequently encountered during the first few days of conventional antidepressant therapy. Treatment with tricyclic imipramine-type compounds has clearly demonstrated this dissociation of effects on mood, inhibition, anxiety, and alertness. (This is the reason for the need to associate sedative psychotropic treatment initially.) The same applies to so-called incisive neuroleptics such as thioproperazine and more recently carpipramine which are disinhibitory at low doses. With sulpiride, this disinhibitory action, described as being predominant at average doses of 400–600 mg, can sometimes reduce the anxiety threshold level.

Anxiety

This Product does not in fact possess univocal activity; in most cases it is described as anxiolytic, sometimes as sedative (but this term *sedation* is very ambiguous as it is not always clear as to whether it applies to sedation of anxiety, an effect on alertness, or a sedating effect on psychotic symptoms such as agitation, hallucinations, delusions . . .). Its mechanism of action is debatable: direct or indirect, secondary to psychodynamic modifications, or related to its antidepressant activity?

Psychotic Syndrome

Finally, the effect on the psychotic nucleus and particularly autism, athymia, and apragmatism is usually favourably compared with that of reference neuroleptics such as chlorpromazine, thioridazine, haloperidol, and perphenazine; sulpiride is then qualified as being antipsychotic and 'resocializing'. It is generally considered as being less hallucinolytic than haloperidol, though it has been observed to produce a more rapid and more satisfactory insight into delusional themes.

Comparative studies against neuroleptics and placebo are greater in number than those comparing activity with antidepressants, and they demonstrate marked improvement of delirious syndromes, schizophrenic or not, the best indications being the forms with predominant psychomotor inhibition. Finally, the lack of biological intolerance even at high doses is an accepted fact, and nearly all authors agree that sulpiride produces few disturbing side-effects, especially an extrapyramidal effect. (The symptoms mentioned most frequently are: dyskinesias, especially of the neck and face, slight tremor, akathisia, tasikinesia, and few syndromes of the akinetohypertonic type.) No sedative effects are observed, and some patients in fact have difficulty in falling asleep for the first few days of treatment. It is in interesting to note that the neuroendocrine effects are very rarely stressed, apart from a few cases of impotence or reduced libido, the amenorrhoea-galactorrhoea syndrome and tendency to gain weight being infrequently mentioned, whereas these latter phenomena appear to us to be the most disturbing.

Overall, all this information leads to discussions and conclusions on the profile of sulpiride which is usually described as an original surprising, upsetting product 'without precedent'. It is true that 10 years ago it was indeed the first molecule, differing from conventional neuroleptics, which confused our most well-established notions concerning the classification of psychotropic agents.

'Our mind is a disorder plus the need to restore order' said Paul Valery. Sulpiride, from 1966 onwards, has thrown disorder into our psychopharmacological concepts: how can a Cartesian mind conceive that a compound could possess two antinomic properties of a neuroleptic and a thymoanaleptic? This is why most authors stress the disinhibitory effects and classify sulpiride among the so-called incisive neuroleptics [on the right in the classification of *Lambert and Reval* (1960)] and do not accept that it possesses effects on mood [e.g., *Volmat* et al., 1969; *Ropert,* 1969, 1971, 1977; *Ropert and Ropert,* 1971, 1972; *Ropert* et al., 1970a, b, 1979; *Loo,* 1979a, b; *Loo and Cuche,* 1979]. This is not always our opinion, however, as will be discussed later.

Clinical Case Reports

Rather than presenting statistical tables, we felt that it would be less daunting and more stimulating to present several cases to illustrate the main properties of sulpiride, taken randomly from hospital or outpatient records between 1970 and 1980. These summaries, anecdotal, will demonstrate the preferential point of impact for each diagnosis and clarify the prescribing procedure and the developmental modalities of treatment with sulpiride.

Case 1: Obsessional Psychoneurosis

In August 1976, Mr. B. was sent to us by a friend. Aged 42 years, embassy attaché, he consulted us as an emergency during a short stay in Paris. Psychasthenic ++, he had anxiety and depression, probably increased by psychoanalysis from 17 to 20 years of age. He doubted everything and had vague persecution ideas. Recurrence of problems for 1 month related to transfer to the USA. He had a fixed attitude and was on multiple neuroleptic treatment (trifluoperazine + amitriptyline 100 mg and diazepam 10 mg for several weeks). His wife was very aggressive and insisted that he had electroconvulsive therapy. Previous treatment was discontinued and we prescribed sulpiride 200 mg twice daily and orphenadrine 200 mg, while continuing amitriptyline and diazepam. Seen 1 week later, just before leaving for the USA, he was smiling, his negative symptoms were greatly improved and his features more mobile. Still anxious about his transfer. 3 months later he returned for a vacation and was much less phobic and depressed. Taking sulpiride 400 mg and orphenadrine 200 mg only. We prescribed large quantities of sulpiride as it is not available in the USA and he stated that 'it is the only medicine that improved me in such a spectacular manner since I became ill 20 years ago'. Seen yearly since then, and is taking sulpiride 400 mg daily only. In spite of a further transfer is perfectly adapted to his work and has almost no further marital problems.

Spectacular and rapid improvement, with no side-effects, in a patient with an obsessional psychoneurosis which had interfered with normal living.

Case 2: Acute Psychosis

A 31-year-old woman, hospitalized for the first time in 1976, with acute anxiety, depression, and delirium, with agitation and crying 3 months after a miscarriage. She was semimute, refused food, and had only slightly structuralized ideas of persecution. Because of the dual nature of the clinical picture, with acute psychosis and severe depressive elements, treatment was immediately started with sulpiride as injections of 2 ampoules morning and midday. Within 2–3 days the patient emerged from her confused and delirious state, contact was improved and anxiety and depression very much less. Still perplexed with some ideas of persecution. Because of a mild akineto-hypertonic syndrome she was prescribed orphenadrine 200 mg and oral treatment started while continuing with diazepam 20 mg prescribed initially for the anxiety. After 2 weeks in hospital the patient was almost cured and was able to leave the hospital taking sulpiride 400 mg twice daily, with orphenadrine and diazepam. The sulpiride dosage was gradually reduced and after 1 year's treatment with a low dose of 100 mg the psychic condition was still very satisfactory. The patient complained of endocrine-type side-effects with a gain in weight of 5 kg and amenorrhoea and galactorrhoea. 3 years later, after a further pregnancy, she developed a similr acute psychotic syndrome (did not follow our advice to continue on a small maintenance dose after delivery). Treated with the same therapy with the same results, and the patient is still taking sulpiride, but has no side-effects as *3 tablets of bromocriptine daily were added* which totally prevented the endocrine effects.

Rapid and spectacular recovery from two acute delirious and depressive psychotic attacks after sulpiride administration. Endocrine side-effects were easily controlled by bromocriptine.

Case 3: Alcoholic Psychosis

Ideas of persecution with a background of a hallucinatory syndrome in a 40-year-old woman with chronic alcoholism for about 15 years. Anxious depressed personality who became worse after haloperidol and then improved several times with sulpiride i.m. 300–400 mg followed by 800 mg orally which was progressively reduced. The patient stopped therapy each time she left the hospital and started to drink again. Pipotiazine L4 was started and she attended for injections, but complained of side-effects of the akinetohypertonic type and especially asthenia, depression, and anxiety. She stopped attending and was admitted several months later because of hallucinations and suicidal thoughts, heavy drinking, and depression. Accepted treatment with sulpiride only (400 mg daily orally) and 1 week later the patient was smiling and relaxed: 'the voices have gone and I don't need to drink', and there were no side-effects.

Improvement of a hallucinatory syndrome due to alcoholism, with depression and anxiety after no improvement with other therapy. No side-effects, long-term treatment was possible.

Case 4: Manic-Depressive Psychosis

A 35-year-old woman was brought to consultation because of severe depression with extreme guilt, self-accusation, ideas of being incurable, insomnia, loss of weight, etc, and symptoms due to neuroleptic therapy with associated antidepressants, hypnotics, and various anxiolytics. Depression had been present for 1 year and appeared related partly to marital and family problems. Sulpiride treatment was commenced with 2 capsules of 50 mg twice daily and orphenadrine for the neuroleptic syndrome. Clorazepate and a hypnotic were also given. She improved within 1 week, was less anxious, spoke of her problems, but was still depressed. After 2 months the patient was much less sad and did not speak of her depressed mood. After 5 months the therapy was reduced to 3 and then to 2 capsules, and after 1 year sulpiride treatment was discontinued because of her normal mood and return to work. A manic episode developed after several months. Her mother then informed us that one of the patient's uncles and a sister were treated for manic-depressive psychosis. Lithium was, therefore, prescribed but was replaced by sulpiride when no improvement occurred. After 1 month there was improvement in the depression.

Very favourable effect on the depressive phases of a manic-depressive psychosis, and antiparkinson therapy prescribed with neuroleptics could be discontinued with sulpiride treatment.

Case 5: Involutional Depression

The patient, a woman, first consulted us 10 years ago because of depression at the age of 60 years. She lacked, self-esteem, felt incapable, and had lost interest in life. 'I am a burden to my children, I cannot do anything, and might just as well die', which was very striking in a woman with an exuberant personality and chronic hypomania. After 1 week on sulpiride (50 mg capsules, 2 twice daily). She was much improved, smiling, and had no side-effects, looking after herself again and had developed insight. Treatment was discontinued after 3 months, and there was almost total recovery. 10 years later, after the death of her husband who she 'mothered because he was very ill', she developed a further attack with severe reactive depression, crying constantly, and was almost completely inactive. The same therapy was started again together with lorazepam because of severe insomnia and anxiety. Improvement was slower, but was much better after 2–3 months, sleeping was better and taking interest in daily activities. Treatment was progressively reduced over 1 year and then stopped. The patient remains well.

Improvement of two episodes of involutional depression, one apparently 'spontaneous', the other 'reactive', without side-effects in an elderly woman with a cyclothymic personality.

These cases are good illustrations of the improvement and consolidation noticed during the initial trials: these psychotic states of various causes all reacted remarkably rapidly, whether they were initial attacks or relapses. Confusion disappeared rapidly if it existed, a feeling of well-being with reduction in anxiety was noted, followed by disappearance of morbid thoughts and suicidal ideas, allowing a more rapid psychotherapeutic approach in all cases.

The problems concerning the mode of administration: increasing doses during the 1st week or high doses initially, and what dosage according to the clinical picture? These are one of the stumbling blocks for sulpiride treatment. Its therapeutic qualities have to be precisely known, both in relation to dosage and psychological examinations of the patient (personality traits – target symptoms to be treated) in order to prescribe adequate therapy.

Our practice is to treat all depressive states with low dosages, associating a benzodiazepine (diazepam, clorazepate, lorazepam) if anxiety and insomnia were severe.

For psychotic syndromes with hallucinations and delusional themes associated with depressed mood, high doses of sulpiride are required initially. Moderate doses are reserved for hebephrenic syndromes and states with predominant psychomotor inhibition.

In summary, therefore, and after 15 years of clinical experience with sulpiride, the following therapeutic regimen is used: (1) thymoanaleptic at low doses, below 400 mg; (2) disinhibitory at moderate doses, from 400 to 600 mg, and (3) antipsychotic at high doses, above 600 mg.

As demonstrated in cases 4 and 5, we never observed mood inversion during therapy of depressive states in cyclothymic personalities. This is an important point, as not only was no manic inversion noted, but also no suicidal tendencies developed (which avoids associated therapy with a sedative neuroleptic), whereas an increase in manic excitation states may occur.

An anxiogenic effect (feeling of tension) may occur at high doses in certain cases, which we interpret as too great a disinhibitory effect or the first signs of an extrapyramidal syndrome. If mild symptoms appear, the addition of low doses of an antiparkinsonian agent causes rapid disappearance of the syndrome.

Sulpiride itself has no atropinic activity. Rare cases of mild hypertension, but no hypotension was observed. Neurological side-effects are rare. As reported in foreign articles, we noted rare cases of facial dyskinesia, but no case of tardive dyskinesia has been observed, as occurs with other neuroleptics.

Sexual and neuroendocrine disturbances have been encountered, however: reduced libido, but mainly amenorrhoea-galactorrhoea which could be easily controlled with bromocriptine.

The final problem, and the most important one for psychotic agents, is the effect on alertness. To our knowledge, and apart from 1 case observed by ourselves and 2 or 3 cases reported by *Bouvier* et al. [1971], sulpiride never provokes somnolence (no sedative effect on alertness). On the contrary, the problems associated with difficulty in falling asleep are related to its noo-analeptic activity when it is prescribed after 4 p.m. [*Borenstein* et al., 1969a, b; 1979a, b; *Mori and Kato,* 1973; *Vivier,* 1973]. Neurophysiological studies have confirmed these clinical observations.

These properties enable sulpiride to be prescribed for effective therapy of alcoholic syndromes and addiction [*Baciocchi* et al., 1970; *Planche and Planche*, 1975] both for acute disorders and during withdrawal therapy (associated with a benzodiazepine if necessary). Sulpiride does not cause dependence.

On the basis of all this information the major contraindications, according to all authors, are states of agitation or paranoia with aggressiveness and multiple complaints where nervous tension and excitation are predominant.

Discussion

Sulpiride has been available to physicians in many countries for more than 10 years, but its two major properties, antipsychotic and thymoanaleptic, are still the subject of a great deal of controversy. Qualified as the 'aspirin of psychiatry' by one author, perhaps it has been too often presented as the universal panacea for all psychiatric and psychosomatic disorders: for many physicians this double polarity has been the cause of confusion, however. Nevertheless, the discovery of new molecules with original chemical formulae and therapeutic properties forces us to revise our psychopharmacological concepts more and more.

As examples we can quote carpipramine, minaprine, amoxapine, and loxapine: 'transition neuroleptics' which have led to new profitable research developments, as has sulpiride, in the pharmacological, biochemical, and neurophysiological fields.

At the same time as psychopharmacological classifications are being reviewed and because of the possibility of prescribing early treatment and continuing for long periods, the conventional nosographic concepts, especially of depressive states, are having to be revised. *Pichot* [1978] clearly outlined the problem when he compared the latest data adopted by the *American Psychiatric Association* [1980]. The dominant idea is that each case should be defined by a series of symptoms (operational criteria) 'in such a way that there is perfect agreement between two psychiatrists examining the same patient and using the same criteria'.

It is surely more pragmatic and scientific to be able to discuss the effects of a compound on target symptoms than on large syndromes. We treat sadness as a symptom, whatever the type of depression. As a further example, *Widlocher and Binoux* [1976], treat psychomotor retardation as the main symptom of depressive states. It is in this context that we approached the delicate problem of whether this effect of sulpiride in reestablishing contact, producing a feeling of well-being, improving social activities, and allowing rapid psychotherapy, is related to a purely disinhibitory or antidepressive action (or both at the same time) and/or an anxiolytic effect. Our early opinion, reinforced by years of experience with sulpiride, has convinced us that it is mainly a question of dosage.

One of the main arguments suggesting that sulpiride has disinhibitory, but not thymoanaleptic activity is that it never causes mood inversion [*De Buck,* 1979a, b; *Deniker,* 1979a, b; *Deniker* et al., 1975, 1979; *Loo,* 1979a, b; *Loo and Cuche,* 1979; *Morel,* 1969; *Ropert,* 1969, 1971, 1977; *Ropert and Ropert,* 1971, 1972; *Ropert* et al., 1970a, b, 1979]. In fact, it can produce manic types of excitation (several cases have been reported) and increase manic syndromes. Furthermore, depression is not always improved by mood inversion or even hypomania. 'True antidepressants' do not always cause mood inversion and moreover, as sulpiride possesses psycholeptic properties, is it not itself its own corrector of mood inversion (as if we prescribe imipramine-levomepromazine or clomipramine-haloperidol for example) [Lille Symp. to be published]. Our opinion is that there are no 'true' or 'false' antidepressants, but conventional and original antidepressants.

What can we learn from experimental studies? The problem was well situated by the initial chemical and pharmacological data obtained: an original molecule, shown by pharmacokinetic studies to be localized mainly in the pituitary in mice and with dissociated neuroleptic properties. A question arises as to whether the metabolites of sulpiride possess antidepressive properties as with loxapine.

Sulpiride definitely affects dopaminergic systems, as shown by many pharmacological tests [*Jenner and Marsden,* 1979], and its antidopaminergic (and not anticholinergic) action can explain its neuroleptic activity, as with other benzamides. This hypothesis is supported by measuring homovanillic acid levels [*Ackenheil* et al., 1979]. Its main effect is on the hypothalamus-pituitary axis, particularly well studied in neuroendocrinology [*Soulairac and Soulairac,* 1979], and this confirms its clinical effects. The action on prolactin secretion (perhaps through the prolactin-inhibiting factor raises a problem: as if it is considered that at so-called antidepressant doses the dopaminergic effect is inversed (dopa plus or related to other hypothetical receptors), how can one explain the even more frequent incidence of endocrine syndromes at low dosages. These are easily controlled by bromocriptine which has confirmed dopaminergic activity.

Nomifensine, a new antidepressant, could possess dopa plus hypoprolactinaemic properties [*Cayreyre,* 1980] and it is known that compounds that increase dopaminergic transmission increase manic states (e.g., Parkinson's disease treated with *L*-dopa). An action on the noradrenergic system has also been demonstrated [*Ackenheil* et al., 1979]. Increased 3-methoxy-4-hydroxyphenylglycol levels after sulpiride strongly suggest an increased turnover of noradrenaline (in comparison with haloperidol and two so-called disinhibitory compounds: carpipramine and clozapine). It would be interesting to know whether there are more or less low 'responding types' to 3-methoxy-4-hydroxyphenylglycol as to imipramine and amitriptyline.

As for an action on 5-hydroxytryptamine (reduction of 5-hydroxy-indoleacetic acid levels in the cerebrospinal fluid in manic-depressive psychoses), modifications in depressed patients have not been clearly demonstrated with benzamides. Neurobiochemical hypotheses are continually the subject of debate because of the large amount of information developed and the contradictions between experimental and clinical data; but there is some agreement between neurophysiological and psychopharmacological data.

Throughout the period of clinical studies with sulpiride we have never observed major modifications in EEG rhythms, either after single or repeated doses or short- or long-term treatment. More particularly, at normal doses of 100–1,000 mg, no convulsant activity or even slow-pointed dysrhythmias, as usually observed with neuroleptics and tricyclic antidepressants, have been observed. Of even greater importance is the fact that during studies of sleep pharmacology [*Borenstein* et al., 1976a, b] we never noted significant variations in the different nocturnal phases from the quantitative point of view: conservation of paradoxical sleep is of particular importance as sulpiride was the only compound out of 24 tested which did not reduce rapid-eye movement; qualitatively, however, sleep appears to be lightened (reduction in stage IV in relation to stages II and III).

Animal studies in free implanted cats [*Borenstein* et al., 1969a, b, 1976a, b; *Borenstein and Cujo,* 1969] have confirmed these data, and during daily administration of relatively high doses (150 mg/kg orally) for long periods, no reduction in paradoxical sleep was noted. Mild behavioural and electrophysiological effects were observed during the waking stage, whereas with the other 25 compounds (psycholeptic or psychoanaleptic) studied, there were always (except for certain benzodiazepines) some disturbances of a major nature in the sleep-awake cycle, and especially reduction in periodic eye movements. Antidepressants such as imipramine and nialamide were particularly studied because of their major activity on paradoxical sleep. More recently we have studied the effects of low doses of sulpiride to determine whether different modifications were produced.

Preliminary studies show that at 50 mg/kg i.m. or orally (i.e. a third of the dose previously administered), there is a reduction in paradoxical sleep; it is difficult to infer that an antidepressant effect is present comparable with that of the tricyclics or monoamine oxidase inhibitors. However, if this argument is developed, there is a reduction in POG with sulpiride (data to be confirmed).

Finally, studies on changes in stimulated hippocampic theta rhythms, both spontaneous and provoked by reticular formation, demonstrate an increase in this base rhythm rate showing limbic vigilance. Direct injections into the posterolateral region of the hypothalamus have confirmed this result and demonstrate a stimulating effect of sulpiride on the hypothalamus and mesolimbus. Only one compound, amphetamine, of the 25 compounds tested increases the rate of hippocampic theta rhythms. Nobody questions that this is a definite noo-analeptic action (by a noradrenergic mechanism?).

It should be recalled, however, that renewed interest has been developed in the use of small doses of amphetamines as antidepressants in man. In contrast, we cannot explain the appearance of hippocampic delta waves after *L*-dopa in the waking state. Still from the neurophysiological point of view, we found no neuroleptic-type cataleptic activity with sulpiride, even at high doses (450 mg/kg orally), and no epileptogenic effect, whereas all these effects were noted at lower doses (50 mg/kg orally) with sultopride which has confirmed neuroleptic activity in man. *Mercier* et al. [1979] have reported similar differential findings in cats and monkeys as far as the effect on vigilance is concerned and after use of the Porsolt's test to compare the disinhibitory and antidepressant effects of the benzamide with other products such as thioproperazine, carpipramine, lithium, and tricyclic antidepressants in mice. Here again, however, the disinhibitory and antidepressant effects at low doses cannot be differentiated.

It is very difficult to compare the different experimental study results, as well as clinical findings, because they vary greatly especially for species employed, dosages, and, more particularly, duration of administration.

Conclusion

What are the final conclusions to be drawn today, when compared with 1967? We can only confirm our original data. Perhaps we can emphasize that we have never observed disabling tardive dyskinesias and that the only disturbing side-effect is the amenorrhoea-galactorrhoea syndrome which can now be perfectly well controlled with bromocriptine. Finally, the absence of atropinic effects enables it to be prescribed for elderly patients.

Sulpiride is the first compound to have produced a breach in the classification of psychotropic agents. Possessing definite antipsychotic properties, it cannot be classed among the neuroleptics, even of the disinhibitory type. Even though we have no reservations at all as to its thymo-analeptic activity, especially at low doses, it has no place either among conventional antidepressants. One must accept its originality and employ its polyvalent therapeutic properties as a function of the symptoms to be treated. Each patient requires the best medication adapted to his needs, at adequate effective doses.

References

Abe, K.: Sulpiride in depressive school phobic children. Psychopharmacologia *43:* 101 (1975).

Abe, K.: Treatment of paranoid depressives and depressive school phobia. Int. Pharmacopsychiat. *14:* 110–113 (1979).

Ackenheil, M.; Fischer, B.; Zander, K.J.; Zimmer, R.: Recherches biochimiques et pharmacologiques sur le mécanisme d'action du sulpiride chez des rats et des patients schizophréniques. Psychol. méd. *11:* 2763–2774 (1979).

American Psychiatric Association: Diagnostic and statistical manual of mental disorders; 3rd ed. (Washington 1980).

Baciocchi, M.; Eppe, J.; Fournier, A.; Gayral, L.; Grandmontagne, O.; Maurel, H.; Roux, S.: Essai clinique du sulpiride. Gaz. méd. Fr. *77:* 3129–3131 (1970).

Borenstein, P.: Expertise clinique du RD 1403 – Sulpiride (1967).

Borenstein, P.; Champion, C.; Cujo, P.; Gekiere, F.; Olivenstein, C.; Kramarz, P.: Un psychotrope original: le sulpiride. Sem. Hôp. Paris *45:* 1301–1314 (1969a).

Borenstein, P.; Cujo, P.: Electro-encéphalographie clinique et substances psychotropes. Sem. Hôp. Paris *45:* 1315–1330 (1969).

Borenstein, P.; Cujo, P.; Champion, C.; Olivenstein, C.: Etude d'un nouveau psychotrope: le sulpiride (1403 RD). I. Méthode et résultats cliniques. Annls méd.-psychol. *2:* 90–99 (1968a).

Borenstein, P.; Cujo, P.; Champion, C.; Olivenstein, C.: Etude d'un nouveau psychotrope: le sulpiride (1403 RD). II. Etude clinique et électro-encéphalographique. Annls méd.-psychol. *2:* 100–107 (1968b).

Borenstein, P.; Cujo, P.; Soret, C.: Pharmacologie du sommeil nocturne de l'homme adulte. Annexe: examen des résultats expérimentaux selon la méthode d'analyse factorielle des correspondances. Sem. Hôp. Paris *52:* 50–79 (1976a).

Borenstein, P.; Gekiere, F.: Etude clinique et électro-encéphalographique de deux *o*-anisamides: le sulpiride et le sultopride. Psychol. méd. *11:* 2833–2836 (1979).

Borenstein, P.; Gekiere, F.; Allegre, G.; Brindeau, F.: Action sur le cycle veille-sommeil de deux dérivés des *o*-anisamides substituées: le sulpiride et le sultopride. Pathol. Biol. *27:* 439–443 (1979a).

Borenstein, P.; Gekiere, F.; Allegre, G.; Brindeau, F.; Cleau, M.: Pharmacologie du cycle veille-sommeil chez le chat implanté libre. Sem. Hôp. Paris *52:* 5–49 (1976b).

Borenstein, P.; Gekiere, F.; Allegre, G.; Brindeau, F.; Cleau, M.: Effets comparés du nialamide, et l'imipramine et du sulpiride sur le sommeil paradoxal du chat implanté libre; in Passonant, Oswalt, Pharmacology of the states of alerteness. Proc. Satellite Symp. 7th Int. Congr. of Pharmacology, Montpellier 1978, pp. 265–267 (Pergamon Press, Oxford 1979b).

Borenstein, P.; Gekiere, F.; Gruet-Masson, J.; Menchikoff, F.; Allegre, G.; Cujo, P.: Etude comparative de l'effet de trois neuroleptiques majeurs sur l'évolution des tracés de veille-sommeil de chats implantés libres. Sem. Hôp. Paris *45:* 1271–1289 (1969b).

Bouvier, C.; Masquin, A.; Amieux, M.-C.: Indications psychiatriques du sulpiride à propos de 60 observations. J. Méd. Lyon *1206:* 519–533 (1971).

Carrere, J.: Etude des effets du sulpiride sur l'état psychique de quarante malades menteux. Annls méd.-psychol. *2:* 560–574 (1968).

Cassano, G.B.; Castrogiovanni, P.; Conti, L.; Bonollo, L.: Sulpiride versus haloperidol in schizophrenia. A double-blind comparative trial. Curr. ther. Res. *17:* 189–201 (1975).

Cayreyre, M.: Alival: indications originales. Etudes en cours. Nouvelles perspectives. Psychol. méd. *12:* 91–98 (1980).

Cicchetti, V.; Gabrielli, F.; Giberti, F.; Muzio, M.: Sulpiride in the treatment of psychoneurosis. Int. Congr. on Neuropsychiatry, Copenhagen 1972. Psychopharmacology suppl. p. 26 (1972).

Collard, J.: Le sulpiride, un neuroleptique singulier, anti-autistique et thymanaleptique. Sem. Hôp. Paris *45:* 3028–3033 (1969).

Collard, J.: Le sulpiride parmi les techniques antidépressives de pointe. Aujourd'hui qu'en dire, qu'en penser? Revue méd. Liège *31:* 135–138 (1976).

Cornet, A.; Grivaux, M.: Recherches phsysiopathologiques sur le sulpiride en gastroentérothérapies de l'autisme. Acta psychiat. belg. *71:* 42–55 (1971).

Cornet, A.; Grivaux, M.: Recherches physiopathologiques sur le sulpiride en gastroentérologie. Soc. méd. Hôp. Paris: *119:* 753–760 (1968).

Cornet, A.; Grivaux, M.: Le sulpiride dans la maladie ulcéreuse. Sem. Hôp. Paris: *45:* 840–861 (1969).

De Buck, R.P.: Thérapeutiques biologiques de la psychose maniaco-dépressive. Encycl. méd.-chir. Psychiatrie, 37221 A–10 (Masson, Paris 1979a).

De Buck, R.P.: Les propriétés antidépressives éventuelles du sulpiride. Psychol. méd. *11:* 2821–2826 (1979b).

Deniker, P.: Nosologie moderne des troubles de l'humeur. Encéphale *5:* 407–416 (1979a).

Deniker, P.: L'encéphale. Coll. Int. sur l'Approche Moderne des Désordres de l'Humeur, Monaco 1979b.

Deniker, P.; Ginestet, D.; Loo, H.: Médicaments psychotropes et mécanismes de régulation de l'humeur. Encéphale *1:* 359–362 (1975).

Deniker, P.; Zarifian, E.; Poirier, M.: General survey and classification of drugs used in france. Revue Psychiat. Univ. Ottawa *4:* 149–155 (1979).

Jenner, P.; Marsden, C.D.: Mécanisme d'action des benzamides substituées. Psychol. méd. *11:* 2775–2790 (1979).

Josserand, F.: Le Dogmatil en gériatrie. J. Méd. chir. prat. *149:* 118–125 (1978).

Lambert, P.; Revol, L.: Classification psychopharmacologique et clinique des différents neuroleptiques. Presse méd. *48:* 1509–1511 (1960).

Loo, H.: Thérapeutique des états dépressifs. Médecine *182:* 19–24 (1979a).

Loo, H.: Effets cliniques des *o*-anisamides. Psychol. méd. *11:* 2839–2840 (1979b).

Loo, H.; Cuche, H.: Classification des antidépresseurs. Encéphale *5:* 591–603 (1979).

Mercier, J.; Scotto, A.M.; Dessaigne, S.: L'action désinhibitrice de certains neuroleptiques – est-elle prévisible expérimentalement? Soc. Biol. *173:* 788–796 (1979).

Morel, P.: Premiers essais cliniques du sulpiride dans un service de psychiatrie générale. Oust méd. *22:* 597–605 (1969).

Mori, T.; Kato, N.: Clinical experience with FK-880 in the treatment of depression and depressive states. Basic Pharmac. Ther. *1:* 59–66 (1973).

Niskanen, P.; Tamminen, T.; Viukari, M.: Sulpiride vs. amitriptyline in the treatment of depression. Curr. ther. res. *17:* 281–284 (1975).

Pichot, P.: Les voies nouvelles de la dépression (Masson, Paris 1978).

Planche, R.; Planche, S.: Sulpiride et toxicomanie. Encéphale *1:* 49–52 (1975).

Ropert, R.: Nouveaux développements dans le domaine des neuroleptiques. Semin Thér. *45:* 291–293 (1969).

Ropert, R.: Originalité thérapeutique en psychiatrie d'un nouveau neuroleptique à indications polyvalentes: le sulpiride (Dogmatil). Sa place parmi les chimiothérapies psychiatriques. XIé Congr. Int. Thér., Barcelone 1971.

Ropert, R.: Anxiété et dépression ne doivent pas être confondues. Quot. Méd. *7:* 20 (1977).

Ropert, R.; Levy, L.; Ropert, M; Coudiere, G.; Weil, D.: Notre expérience clinique du sulpiride en psychiatrie. Sem. Hôp. Paris *46:* 109–116 (1970a).

Ropert, R.; Levy, L.; Ropert, M.; Hullin, R.: Réflexions sur le mode d'action du sulpiride et sur sa place parmi les psychotropes. C.r. 68é Congr. Psychiat. Neurol. Lang. Fr. Milan 1970, pp. 805–813 (Masson, Paris 1970b).

Ropert, R.; Petitjean, S.; Caillard, S.: Les problèmes thérapeutiques posés par les chimiothérapies de la dépression chez le sujet âgé. Gaz. méd. Fr. *86:* 1063–1069 (1979).

Ropert, R.; Ropert, M.: Rôle du sulpiride dans le traitement des syndromes dépressifs. Infs psychiat. *47:* 477–480 (1971).

Ropert, R.; Ropert, M.: Situation du sulpiride (Dogmatil) dans la gamme des antidépressifs (Table ronde 1971). Bordeaux méd. *5:* 127–134 (1972).

Soulairac, A.; Soulairac, M.-L.: Les effets des anisamides sur les résponses sexuelles endocriniennes et comportementales (aspects cliniques et comportementaux). Psychol. méd. *11:* 2795–2805 (1979).

Vivier, L.M.: A qualitative assessment of the clinical use of a new psychotropic preparation, sulpiride. A preliminary communication. Med. Proc. *19:* 28–30 (1973).

Volmat, R.; Allers, G.; Vittouris, N.: Le sulpiride et ses associations, ses résultats sur les troubles de l'humeur et les psychoses de la personnalité. J. Méd. Besançon *5:* 121–132 (1969).

Widlocher, D.; Binoux, F.: Dépressions. Revue Psychol. appl. *28:* 39 (1976).

Yura, R.; Kato, Y.: A double-blind comparative study of the effects of sulpiride and imipramine on depression. Clin. Psychiat. *18:* 89–102 (1976).

P. Borenstein, MD, Hôpitaux Psychiatrique de la Seine, 71, quai Branly,
F–75007 Paris (France)

Adv. biol. Psychiat., vol. 7, pp. 154–165 (Karger, Basel 1981)

Sulpiride in Depressive Illness

A Controlled Clinical Comparison of Sulpiride and Amitriptyline in Patients with Primary Depressive Illness

Mansel Aylward, Jeffrey Maddock, Peter M. Dewland, Peter A. Lewis

Department of Clinical Research and Experimental Medicine,
Simbec Research Laboratories, Cefn Coed, Merthyr Tydfil, Mid Glamorgan, UK

Introduction

Sulpiride (N-{1-ethyl-2-pyrrolidinyl-methyl}-2-methoxy-5-sulfamoyl-benzamide) has emerged as a new major tranquillizer belonging to the class of compounds known as *o*-anisamides [1]. The drug is rapidly absorbed from the human gastro-intestinal tract [2] and its plasma elimination half-life is of the order of 8 h. Moreover, a single dose is almost completely eliminated within 24 h following administration.

A therapeutic efficacy comparable to chlopromazine, haloperidol and perphenazine has been demonstrated in controlled studies of patients with acute and chronic schizophrenia [3–5]. *Niskanen* et al. [6] have demonstrated the antidepressant effects of sulpiride in comparison with amitriptyline in patients suffering from either psychotic or neurotic depression. In particular, these authors noted a superiority of sulpiride over amitriptyline in the relief of psychomotor retardation.

As with other active psychotropic agents (particularly the phenothiazines), sulpiride modifies certain endocrine functions; notably, it stimulates prolactin secretion and suppresses growth hormone. These hormonal effects have been a cause of concern in respect of a possible tumorigenic potential but the stimulation of prolactin secretion may also be tentatively proposed as an indication of the mode of action of sulpiride by its documented antidopamine effects [7].

The study reported here investigated the antidepressant activity of sulpiride in comparison with amitriptyline in a double-blind between-patient (parallel) clinical trial in patients with primary depressive illness. The study also investigated several laboratory biochemical and endocrinological parameters and documented the nature and frequency of untoward effects encountered on the two treatments.

Patients and Methods

Potential participants were selected from patients who had been referred to the out-patient clinic by their general practitioners for assessment and treatment of a primary depressive illness. Informed consent was obtained from all patients who were admitted to the study. Patients under the age of 18 or over 65 years were excluded as were pregnant women, those with a concomitant illness which might have interfered with the conduct and/or results of the study and those who exhibited any known contra-indications to the administration of tricyclic antidepressants or benzamide drugs. Patients with a history of alcohol and/or drug abuse, those with organic brain disease and patients who had received monoamine oxidase inhibitors during the preceding 2 months or recent electro-convulsive therapy were also excluded from participation.

Additionally, all patients fulfilled the following further criteria before enrolment to the study: the patient's condition had persisted for at least 2 weeks prior to admission to the study but the current phase of the depressive illness had not persisted for more than 2 months; a minimal score of 25 on the Hamilton Rating Scale (HRS) for depression [8], of which the sub-score for suicide was less than 3 points, and antidepressant drugs had not been taken during the preceding 2 weeks.

The study was undertaken in accordance with the Provisions of the Declaration of Helsinki and the Tokyo Revisions.

Following a comprehensive clinical and laboratory examination and psychiatric interview to document the symptomatology of the disorder, previous history and the existence of any family and social adverse circumstances, a depressive diagnosis was confirmed, based upon which the patient was categorized as suffering from either depressive psychosis (ICD 296.2) or depressive neurosis (ICD 300.4) in keeping with the eighth edition of the International Classification of Diseases and the British Glossary to it [9]. The sub-group of reactive depressive psychosis (ICD 298.0) was not used; thus patients with a depressive psychosis with a symptomatology similar to that described under manic depressive psychosis, depressed type (ICD 296.2) were assigned to ICD category 296.2.

Following stratification for age and sex, patients were randomly assigned either to sulpiride or to amitriptyline. The double-dummy technique was used to ensure that the double-blind nature of the study was preserved.

Patients assigned to the sulpiride treatment received initially one sulpiride tabled 200 mg twice daily plus one dummy amitriptyline tablet (25 mg) three times daily, and those assigned to the amitriptyline treatment received one amitriptyline tablet 25 mg, three times daily, plus one dummy sulpiride (200 mg) twice daily.

It was felt that the use of a standardized dosage reduced the chances of demonstrating the difference in situations where two antidepressants of potentially unequal potency were being tested; accordingly a flexible dosage schedule was used in which the dose was gradually increased until the patient showed a considerable improvement and then continued at a maintenance level until completion of the investigation. In general, the dose level was not increased during the first 5 days.

The daily dosages ranged from 75 to 200 mg for amitriptyline and from 400 to 1,000 mg for sulpiride.

Nitrazepam up to a maximum of 10 mg at night was permitted if the patients had already been in receipt of this drug for at least 4 weeks before admission into the study. Diazepam was also continued up to a maximum daily dose of 15 mg if patients had been in receipt of this drug for at least 4 weeks before admission.

Table I. Basic data and diagnostic categories on patients assigned to the two treatment groups

	Sulpiride	Amitriptyline
Number of patients (including withdrawals)	25[1]	25[2]
Number of female patients	14	16
Age (mean ± SD), years	45.5 ± 9,48	46.5 ± 9.56
Endogenous depression	11	10
Depressive neurosis	14	15

[1] Patients subsequently withdrawn = 4
[2] Patients subsequently withdrawn = 5

Assessments

The degree of clinical psychopathology was expressed in ratings using the HRS for depression, the Beck Self-Rating Scale and a 15-point Clinical Global Impression Scale and a Visual Analogue Scale (VAS) for depression. These were completed weekly for the first 4 weeks of the study and then at 6 weeks which marked the end of the trial.

Laboratory testing included haematology, selected biochemical tests of renal and hepatic function, protein electrophoresis, total and free plasma tryptophan, whole blood serotonin and serum prolactin levels. These were measured initially and at each visit at the same time every day throughout the study. In female patients plasma gonadal steroid and protein pituitary hormones were measured initially and at each visit. A side effect checklist was completed before admission to the study but during the study period itself direct questions were not asked to elicit side effects.

Results

A period of 13 months was necessary to enroll a group of 50 patients who fulfilled the entry criteria for this study. Of the 50 patients admitted to the study, 9 were subsequently withdrawn: one was referred for electro-convulsive therapy, one failed to adhere to the assigned drug regimen, and 7 exhibited adverse clinical experiences which necessitated withdrawal. These are described later.

Details of age and sex of the patients and their distribution by diagnosis into the two treatment groups are given in table I.

The mean scores on the HRS for the sulpiride and amitriptyline groups immediately before treatment, at each week during the first 4 weeks, and at the end of the study (week 6) are given in table II. The mean scores were similar and

Table II. Total HRS scores (depression) in the two treatment groups at each study assessment (means ± SD)

Group	Week 0	Week 1	Week 2	Week 3	Week 4	Week 6
Sulpiride	26.70 ± 2.04	24.52 ± 2.89	22.72 ± 3.87	15.63 ± 4.78	11.87 ± 5.44	8.95 ± 3.67
Amitriptyline	25.40 ± 4.60	24.71 ± 3.37	21.50 ± 3.22	11.81 ± 3.47[1]	9.75 ± 3.04	8.05 ± 3.28

[1] Significant difference in favour of amitriptyline (F = 9.14).

Table III. The first assessment (week) at which specific symptoms showed significant improvement (p < 0,05)

Assessment	Sulpiride	Amitriptyline
After 1 week	impairment of concentration anxiety (somatic)	anxiety (psychic)
After 2 weeks	work and interests insomnia (middle) suicide guilt	suicide anxiety (somatic)
After 3 weeks	retardation depressed mood irritability	insomnia (delayed) insomnia (middle) depressed mood hypochondriases work and interest
After 4 weeks	anxiety (psychic) agitation insight insomnia (delayed)	agitation self-pity diurnal variation impairment of concentration
After 6 weeks	diurnal variation hypochondriases self-pity	insomnia (initial) insight guilt

the differences not significantly different apart from week 3 scores when the result was in favour of the amitriptyline group ($p < 0.05$). Analysis of covariance indicated that the changes in the total HRS scores was not significantly different for the two groups when final scores were compared with initial scores.

Since assessment on the HRS employs the scoring system graded either 0–4 or 0–2; in each case the higher the score the more severe the symptom, there are

Table IV. Clinical global evaluation scores in the two treatment groups at each study assessment (means ± SD)

Group	Week 0	Week 1	Week 2	Week 3	Week 4	Week 6
Amitriptyline	3.96 ±0.61	3.80 ±0.50	3.66* ±0.70	2.77** ±0.76	1.95** ±0.68	1.70** ±0.65
Sulpiride	4.20 ±0.50	4.04 ±0.53	3.56** ±0.71	2.29** ±0.90	2.13** ±0.96	1.66** ±0.57

Significant reductions compared with pre-treatment values: *p < 0.01; **p H 0.001.

Table V. VAS Scores (mms) for depression (means ± SD) for the two treatment groups

Group	Week 0	Week 1	Week 2	Week 3	Week 4	Week 6
Sulpiride	56.10 ±14.52	54.00 ±13.30	41.60 ±15.05	32.30 ±12.23	24.80 ±12.33	22.42 ± 9.14
vs.						
amitriptyline	52.10 ±11.67	51.80 ±12.16	40.90 ±13.51	26.30 ± 6.27	27.00 ±13.20	23.50 ± 7.61
t	1.07	1.05	0.81	2.02	−0.45	−0.40
d.f.	48	48	45	43	41	39
p	>0.1	>0.1	>0.1	0.05	>0.1	>0.1

thus very few possible outcomes for the individual scores. Employing the sign test for the median, the sub-scores were statistically analysed. If there were no differences in the scores from baseline to time of assessment these were included in the 'no-improvement' group. The results of analysis are shown in table III and these indicate that at each week of assessment an appreciable number of patients show improvement. Notably, impairment of concentration and indices of psychomotor retardation improved earlier in those in receipt of sulpiride. With amitriptyline, however, anxiety (psychic and somatic) and sleep disturbances showed early responses to the treatment.

The sub-scores for depressed mood, guilt and psychomotor retardation on the HRS were examined separately using analyses of variance and analyses of covariance.

The reduction in total specified sub-scores was significantly greater on sulpiride than on amitriptyline at week 4. However, by the end of the study there were no significant differences between the two groups with respect to total subscores for these items.

Table VI. Side effects encountered during the trial in the two treatment groups

	Sulpiride	Amitriptyline
Drowsiness	4	6
Dry mouth	2	13
Blurred vision	1	12
Constipation	0	10
Insomnia	3	6
Nausea	7	3
Tachycardia	4	6
Akathisia	3	0
Diarrhoea	2	1
Tremor	8	6
Sweating	4	7
Dizziness	9	8
Apprehension	7	6
Restlessness	5	3
Total number of side effects	59	87
Number of patients with side effects	14 (56%)	19 (76%)
Number of patients withdrawn due to side effects	3 (12%)	4 (16%)
Number withdrawn due to lack of Effect	1 (4%)	1 (4%)

The results of the clinical global evaluation scores (table IV) strongly supported the clinical findings of improvement. Magnitude of improvement was greater on sulpiride than on amitriptyline after the 2 week of treatment.

Group VAS scores for depressed mood at each visit (table V) indicated that, other than at week 3, differences between the groups were not statistically different. At week 3 there was a significant differences in favour of amitriptyline ($p = 0.05$). Clearly, mean VAS scores were reduced progressively in both treatment groups as the study progressed.

Side Effects

The incidence of side effects, particularly as regards anti-cholinergic effects, was significantly greater among those on amitriptyline, whereas 'extra-pyramidal side effects' were experienced only by those in receipt of sulpiride. 4 patients were withdrawn from amitriptyline treatment because of 'anti-cholinergic side effects' which were unacceptable to them, whereas 3 patients exhibiting akathisia were withdrawn from sulpiride.

The frequency and nature of side effects encountered on the two treatments are given in table VI. Spontaneously reported or clinically evident side effects

Table VII. Laboratory data on serum prolactiin, whole blood serotonin and plasma total, free and percent bound *L*-tryptophan in two groups of patients with depressive illnesses (means ± SD)

	Sulpiride					Amitriptyline				
	week −4	week 0	week 2	week 4	week 6	week −4	week 0	week 2	week 4	week 6
Prolactin, mU/ml	282.0	266.3	1173.1**	1020.2**	950.2*	263.2	279.1	692.9*	816.4**	844.8**
	± 07.7	±103.1	± 483.7	± 353.2	±315.4	±114.4	±109.8	±198.5	±203.7	±229.1
Whole blood 5−HT	118.7	109.0	112.2	108.7	117.3	109.2	106.8	119.2	112.0	113.7
ng/ml	± 27.8	± 21.8	± 24.3	± 21.4	± 25.9	± 25.0	± 21.3	± 26.0	± 20.2	± 23.3
Bound *L*-Trp, %	93.6	93.6	90.9	88.9	89.0	94.4	90.4	92.6	90.0	88.7
	± 3.15	± 3.77	± 3.80	± 2.76	± 3.46	± 3.24	± 3.03	± 3.91	± 3.16	± 3.13
Free *L*-Trp, μg/ml	0.91	0.93	0.97	1.38*	1.52*	0.94	0.98	0.99	0.99	1.26
	± 0.25	± 0.27	± 0.31	± 0.60	± 0.71	± 0.30	± 0.29	± 0.41	± 0.40	± 0.82
Total *L*-Trp, μg/ml	14.52	14.48	14.24	14.39	14.37	14.70	14.51	14.30	14.42	14.39
	± 0.61	± 0.59	± 0.70	± 0.56	± 0.66	± 0.70	± 0.63	± 0.66	± 0.62	± 0.71

Significant: $*p < 0.05$; $**p < 0.001$.

Table VIII. Results of plasma gonadal steroid hormones and protein gonadotrophin concentrations in female subjects in the two treatment groups (means ± SD)

	Amitriptyline group					Sulpiride group				
Hormone concentration	week –4 (16)	week 0 (16)	week 2 (16)	week 4 (13)	week 6 (13)	week –4 (14)	week 0 (14)	week 2 (14)	week 4 (13)	week 6 (12)
Plasma 17β oestradiol, pg/ml	119.0 ± 80.2	139.9 ± 101.4	93.7 ± 64.5	106.8 ± 83.4	142.3 ± 89.0	128.0 ± 75.4	130.6 ± 99.3	112.6 ± 59.6	91.5 ± 62.3	145.1 ± 85.6
Plasma oestrone, pg/ml	294.0 ± 138.1	198.8 ± 122.0	298.6 ± 140.2	189.9 ± 90.9	278.0 ± 149.6	273.2 ± 127.2	209.1 ± 115.9	274.6 ± 137.6	197.2 ± 81.8	250.0 ± 152.3
Plasma progesterone, mmol/l	3.8 ± 4.1	5.9 ± 5.2	5.4 ± 5.3	5.22 ± 4.91	5.57 ± 4.97	5.7 ± 6.4	5.0 ± 4.9	5.83 ± 5.2	5.1 ± 4.16	6.49 ± 5.82
Serum LH, IU/l	52.9 ± 36.7	46.6 ± 35.9	55.9 ± 39.4	59.0 ± 38.9	49.2 ± 35.2	53.2 ± 51.6	60.7 ± 56.9	63.9 ± 66.5	69.7 ± 70.2	69.8 ± 72.2
Serum FSH, IU/l	78.1 ± 80.3	71.4 ± 77.6	71.3 ± 71.9	83.8 ± 85.4	84.3 ± 88.7	69.1 ± 80.9	72.5 ± 66.7	72.0 ± 80.4	81.3 ± 79.6	76.9 ± 87.4

Figures in parentheses are numbers of females.

occurred in 14 patients (56%) on sulpiride and in 19 patients (76%) on amitriptyline. Although the total number of side effects was significantly greater on amitriptyline, the occurrence of nausea and restlessness was decidedly greater on sulpiride.

Serum Prolactin, Whole Blood Serotonin and Plasma Tryptophan Concentrations

In the sulpiride group there were dramatic and highly significant increases in serum prolactin concentrations at the 2nd week of administration (table VII), which were maintained throughout the remaining weeks of the study though values at the 4th and 6th weeks of the study suggested a general decrease in prolactin levels. Significant increases over baseline serum prolactin levels were also observed in the patients receiving amitriptyline; but the magnitude of increase in serum prolactin on amitriptyline was decidedly less than that observed on sulpiride.

Whole blood serotonin concentrations were not appreciably modified by either treatment, and neither total nor protein-bound plasma *L*-tryptophan levels were significantly affected (table VII). However, free plasma *L*-tryptophan concentrations showed a significant increase at weeks 4 and 6 only in patients on sulpiride ($p < 0.05$).

Plasma Gonadal Steroid and Pituitary Gonadotrophin Levels

The data obtained in the female patients are shown in table VIII. Rhythmic changes in accord with the follicular and luteal phases of the menstrual cycle were unaffected by drug administration in the pre-menopausal groups of women. In post-menopausal patients there were no appreciable changes in the levels of the plasma gonadal steroid hormones.

Results for serum levels of LH and FSH are also given in table VIII. In premenopausal women the phasic changes associated with the menstrual cycle appeared unaffected; however, in the post-menopausal patients on sulpiride there was a tendency towards an increase in serum FSH at weeks 4 and 6 but these changes did not reach statistical significance.

Haematology and Blood Chemistry

No significant changes were encountered in haematological indices or in routine tests of blood chemistry.

Discussion

We have compared the therapeutic efficacy and frequency and nature of side effects in patients with primary depressive illness who were treated either with sulpiride or with amitriptyline over a 6-week period in a controlled, double-blind, parallel clinical trial.

Amitriptyline is a tricyclic drug of established antidepressant value in the treatment of depressive disorders whereas sulpiride is a new major tranquillizer belonging to the benzamides which is characterized as a thymo-analeptic neuroleptic agent. The data presented here confirmed the previous findings of *Niskanen* et al. [6] which concluded that sulpiride was more effective than amitriptyline in improving certain depressive symptoms – notably psychomotor retardation.

The most commonly noted undesirable side effects reported with sulpiride have been parkinsonian-like effects which have been reversible with discontinuation of the drug [10]. In our study extra-pyramidal side effects were seen only in patients receiving sulpiride and in all 6 patients so affected the nature of the unwanted experience was that of akathisia which led to the withdrawal of 3 patients. The investigation of sulpiride in experimental animal models [1, 11] has shown only marginal effects on the autonomic nervous system and cardiovascular mechanisms which are mirrored in the very low incidence of associated autonomic disturbances during its use in clinical practice [11, 12]. Our findings indicate no appreciable differences in terms of antidepressant efficacy between amitriptyline and sulpiride. Moreover, patients treated with sulpiride showed a rapid improvement in impairment of concentration and on indices of psychomotor retardation. With both drugs there was greater improvement among patients affected with depressive neurosis (ICD 300.4) than among those with depressive psychosis (ICD 296.2). However, the differences in the magnitude of response between patients in the two diagnostic categories provide insufficient evidence to conclude that either of the drugs was more suited for the treatment of patients classified as either endogenous of neurotic depressives despite the weight of evidence which implies that psychotic depressives respond better to tricyclic antidepressants [13]. An improving effect of hospital admission on neurotic depressives does not cloud the issue in our study undertaken in outpatients. For ethical reasons, placebo treatment was not included in this study; consequently, there can be no estimate of the patients who may have responded without drug treatment. Comparison of the data here with the work of *Ball and Kiloh* [14] in out-patients of whom around 20% both of reactive and endogenous patients improved on placebo, would suggest that the response rates achieved in our study were greatly in excess of those expected to respond well to placebo. Several investigators have found no significant differences between tricyclic antidepressants and phenothiazines in the treatment of depressives, and no interactions with the number of history and symptom predictors, including diagnoses of endogenous and neurotic depression. The assessment of clinical improvement, the selection of a homogeneous group of patients, and allowance for the wide variability in individual response to drug therapy are but a few of the problems which prevent the drawing of any firm conclusions from the results of the study described here. Nevertheless, the quality and quantity of responses documented supported by the results of several studies [3–6] which report the rapidly effective antipsychotic and antidepressant activities of sulpiride provide reasonable evidence to justify the role of sulpiride as a potentially valuable neuroleptic drug with definite antidepressant activities without an unacceptable level of associated side effects.

The data reported here confirm that sulpiride increases prolactin concentrations much more appreciably than amitriptyline. Whether the stimulation of prolactin secretion is in part consequent to the antagonism of dopamine receptors which regulate prolàctin release and so offers an indication of the mode of action of sulpiride and other centrally acting compounds known to antagonize dopaminergic neurones or whether stimulated prolactin secretion is unrelated to the beneficial therapeutic response, or even a potentially harmful accompaniment remains to be clarified. In female patients participating in the study there was no evidence to suggest that either drug affected the pattern or amounts of circulating gonadal steroids and pituitary gonadotrophins.

References

1 Justin-Besancon, L.; Thominet, M.; Laville, C.; Margarit, J.: Constitution chimique et propriétés biologiques de sulpiride. Presse méd. *76:* 86 (1968).

2 Imondi, A.R.; Alam, A.S.; Brennon, J.J.; Hagerman, L.M.: Metabolism of sulpiride in man and rhesus monkeys. Archs int. Pharmacodyn. Thér. *232:* 79–91 (1978).

3 Acevedo, J.C.; Elias, L.V.; Ruiz, J.G.: Double-blind evaluation of the comparative effectiveness of sulpiride and chlorpromazine in treatment of schizophrenia. Congr. Mondial de Psychiatrie, Mexico 1971.

4 Ishimaru, T.; Kubo, S.; Ishikawa, H.; Jitsuiki, S.; Kawamura, T.; Kimura, N.; Kodama, H.; Masuda, K.; Miyake, Y.; Nomura, S.; Sasaki, T.; Schimonaga, K.; Tsukue, I.; Asada, S.: Clinical evaluation on therapeutic effects of sulpiride on schizophrenia by double-blind controlled test. Hiroshima. J. med. Sci. *19:* 131–154 (1971).

5 Bellano, L.E.; Fernandez, H.: Investigations 'double essai aveugle' du sulpiride sur des malades psychatiques. Evaluation psychiatrique et psychologique clinique. Minerva psyquiat. argent. *1:* 29–50 (1972).

6 Niskanen, P.; Tamminen, T.; Viukari, M.: Sulpiride versus amitryptyline in the treatment of depression. Curr. ther. Res. *17:* 281 (1975).

7 Scatton, B.; Bischoff, S.; Dedek, J.; Korf, J.: Regional effects of neuroleptics on dopamine metabolism and dopamine-sensitive adenylate cyclase activity. J. Pharmac. *44:* 287 (1977).

8 Hamilton, M.: A rating scale for depressions. J. Neurol. Neurosurg. Psychiat. *23:* 56 (1960).

9 International classification of diseases. British Glossary (1968).

10 Carrère, J.: A study of the effect of 1403rd (sulpiride) on the psychic condition of 43 mental patients. Annls méd.-psychol. *2:* 560 (1968).

11 Fuxe, K.; Ogren, S.-O.; Fredholm, B.; Agnati, L.; Hokfelt, T.; Perez de la Mora, M.:

Possibilities of a differential blockade of central monoamine receptors. Symposium Bel-Air rhinencéphale neurotransmetteurs et psychoses, Genève 1976.
12 Martin, A.; Masson, H.M.; Thouvenot, J.: L'électrosplanchographie dans les anorexies mentales et les dépressions. Sem. Hôp. Paris *46:* 54 (1970).
13 Paykel, E.S.: Predictions of treatment response; in Paykel, Coppen, Psychopharmacology of affective disorders; pp. 193–220 (Oxford University Press, London 1979).
14 Ball, J.R.; Kiloh, L.G.: A controlled trial of imipramine in treatment of depressive states. Br. med. J. *ii:* 52 (1959).

M. Aylward, MD, Department of Clinical Research and Experimental Medicine,
Simbec Research Laboratories, Cefn Coed, Merthyr Tydfil,
Mid Glamorgan CF48 2PH (UK)

Adv. biol. Psychiat., vol. 7, pp. 166–172 (Karger, Basel 1981)

Use of Psychotropics in Childhood (with Special Reference to Sulpiride)

Gerhardt Nissen

Klinik für Kinder- und Jugendpsychiatrie, Würzburg, BRD

In spite of the substantial differences existing between its French, British and German variants, *European* child psychiatry has always been based on a polyaetiological nosology. This may be a platitude, but sometimes the public and even pedagogues, sociologists, psychologists and a few psychiatrists seem to believe that the opposite is the case, and that all psychic disorders and diseases of children and adolescents are exclusively or at least predominantly due to psychic factors. Needless to say that this is not true.

We could leave this subject if the above remarks helped to abolish the detrimental effects of such attitudes on the treatment of children exhibiting psychic problems, but this obiously does not happen.

Some 30 years ago, in child psychiatry the pre-scientific term 'school and educational problems' was replaced by the not less questionable term 'behaviour disorders'. While educational measures appeared to be the only possible treatment for the correction of 'school and educational problems', the methods used to eliminate 'behaviour disorders' had to change the child's behaviour. According to the main thesis put forward by the US Child Guidance Movements, which had a considerable influence on the educational advisory boards of post-war Germany, behaviour disorders in childhood are primarily due to environmental factors, i.e., to the family or community. As a result, an essential requirement for the correction of such disorders was supposed to be a change in the behaviour of the child or the environment, or preferably of both.

As a matter of fact, behaviour therapy, which is primarily directed at the symptoms, and the more causal psychotherapy have gained an important place in child and adolescent psychiatry. Nevertheless, it should be kept in mind that the disturbed behaviour is not really an illness but a particularly striking symptom, an epiphenomenon like the exanthem associated with an infectious disease, such as measles or scarlet fever.

Like a pressure gage, behaviour disorders indicate the presence of a disturbed development of personality. Hence, they cannot be improved or abolished by repairing the pressure gage. Psychotherapy and behaviour therapy also have indications and contraindications, such as psychoses of childhood, to take but one example. This would appear obious, but for various reasons it is often forgotten or repressed. However, an inevitable consequence of such repression is the discrepancy between the personnel and time spent on psychotherapeutic measures and procedures of behaviour therapy, and the poor results or failure of such treatments. Thus, the euphoric expectations regarding therapeutic possibilities should be revised to avoid paediatricians and social pedagogues merely accepting their failure with resignation. Latent antipsychiatric trends, which are sometimes even encountered in child and adolescent psychiatry, are reflected by an oversimplifying bias and naive paedophilia trying to avoid any diagnosis associated with a presumptively unfavourable prognosis, rather than by memorials and resolutions. However, this precludes effective treatment, above all the use of appropriate psychotropics, which – for other reasons – is controversial in this period of life.

The experience of child psychiatrists has clearly shown that *psychotropics* are used occasionally or frequently by nearly all physicians in the treatment of children. It would appear, therefore, that these drugs may be effective in quite a number of psychic disorders of young patients. However, their use often presents considerable difficulties, since many physicians are not familiar with the diagnosis of psychic diseases in childhood. This would also apply to the selection of suitable drugs, their dosage and the detection and correction of side effects, e.g. in school children. Many physicians simply want to be told *which* drugs are *most appropriate* for *which* psychopathologic conditions. Unfortunately, such requests cannot always be complied with. From the standpoint of child psychiatry, psychotropics cannot be regarded as an especially suitable means to abolish psychopathologic disorders or illnesses in childhood. This is due to the fact that the pharmakokinetics, effectiveness and long-term action of these drugs in that period of life are far from being elucidated. As a matter of fact, it is difficult to obtain sufficiently large groups of children of the same age and stage of development, exhibiting similar symptoms. Moreover, our knowledge of the psychopathology of childhood has been derived from that of adult subjects, and not vice versa. This is also true of psychopharmacology. As the psyche of children cannot be explained by the psyche of adult individuals, the effects psychotropics produce in adults cannot be assumed to be reproduced in young patients.

In *Europe* psychotropic drugs have always been employed with great caution in children. As opposed to various reviews published recently [*Werry,* 1978; *Wiener,* 1977; *Mendlewicz and van Praag,* 1978; *Gittelman-Klein,* 1975], no such reports on paediatric psychopharmacology are available in the German language, with the exception of some incomplete communications [*Nissen,* 1978]. Nor is there any documentation showing the correlations between typical target symptoms [*Freyhan,* 1957] and specific psychotropic agents. Based on our experi-

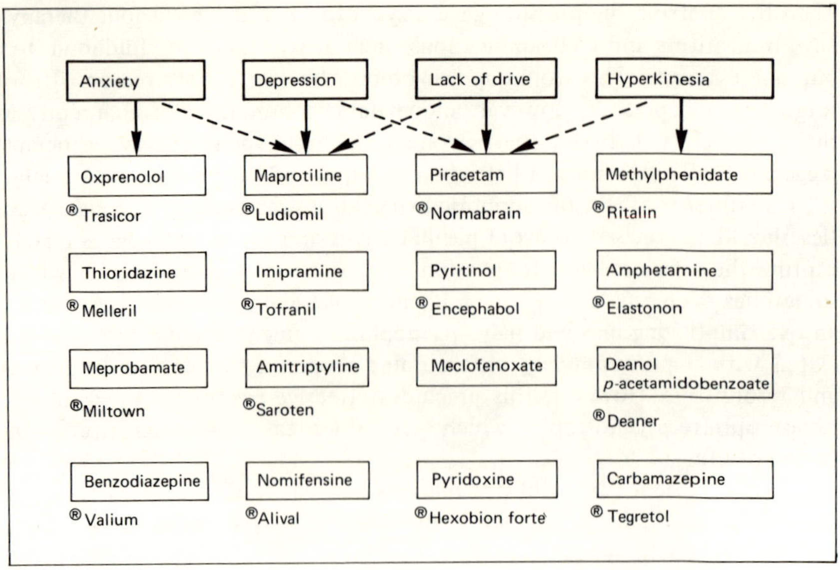

Fig. 1. Target symptoms encountered in children with psychic disorders and drugs indicated in their treatment.

ence, we have formed *four syndrome-like groups* of children by means of a descriptive typological differentiation of psychic disorders [*Nissen,* 1976, 1977, 1978]. In these groups, certain psychotropics were found to produce satisfactory results more frequently. The following target symptoms of psychic abnormalities in childhood were chosen: (1) anxiety; (2) depression; (3) lack of drive; (4) hyperactivity (fig. 1).

In the search for well-tolerated psychotropic compounds providing a wide spectrum of effectiveness and producing rare of minor side effects, we have studied *sulpiride.* According to the reports that have been published on this subject, sulpiride has been used in young people for the treatment of the following disorders: school phobia [*Abe,* 1975; *Lonchamp* et al., 1971; *Picot and Pincon,* 1970], behaviour disorders of childhood [*Rothenberger,* 1976], autism [*Miller and Wallis,* 1979], underweight [*Desbuquois* et al., 1969], learning difficulties [*Puyelo* et al., 1970], neurotic disorders [*Kohler,* et al., 1971], retardation [*Lang and Le Strat,* 1971], psychotic disturbances in childhood and adolescence [*Duche and Delagrange,* 1970] and pre-psychotic syndromes [*Puyelo* et al., 1970]. *Rothenberger* [1976] employed the drug in special indications, such as lack of appetite, playing and learning difficulties, inhibition of working, and autistic behaviour.

A simple *clinical study,* primarily based on carefully compiled, detailed psychopathological data, was performed. Inquiries were made before starting treatment and repeated after 2 and 4 weeks. The parents and, whenever possible, the teachers were questioned regarding the assessment of the patient. All data were entered in appropriate inquiry forms and statistically evaluated.

The following *study population* was selected: anxious, dysphoric and depressed children; isolated, contactless and overadjusted children; children with lack of drive and concentration.

The present study was designed to find out whether sulpiride was capable of producing the following effects: favourable influence upon *anxious, dysphoric and depressive* symptoms in children; improvement of social integration by *increasing* contact and drive; amelioration of *lack of drive and concentration* associated with learning difficulties, and thus indirect improvement of the learning ability; improvement or aggravation of certain *psychosomatic symptoms,* such as thumb-sucking, nail-biting, motor stereotypies, mutism, distrubed sleep, nocturnal anxiety, nocturnal head-rocking, dizziness, unfounded weeping, psychogenic attacks, headache, loss of appetite, abdominal pain, vomiting, daintiness, enuresis, encopresis, smearing with faecal material, genital manipulations, running away.

Furthermore, we tried to settle the following questions: Can the institution and performance of psychotherapeutical procedures be facilitated by the administration of Sulpiride? Is the dosage of the drug suitable for young people? Does the drug produce undesirable side-effects?

Patients

The study was done on children and adolescents with psychic problems on an out-patient basis. They were selected in accordance with a depression scale for children and adolescents [*Nissen,* 1971], which was partly complemented by *Kovac and Beck's* [1977] *depression questionnaire,* translated and modified by one of us.

The study population included *30 children.* The trial was performed on 27 subjects (16 boys and 11 girls), the majority being between 7 and 12 years of age. The mean age of the boys was 10.06 (± 3.62) years and that of the girls 11.09 (± 3.11) years. This difference was not statistically significant. The high incidence of such disorders in the medium grades is a common occurrence in the practice of a child psychiatrist. The sex distribution did not correspond exactly to the worldwide 2:1 ratio of psychic abnormalities in favour of the males. The determination of mean values for height and weight was found to be useless, since we were dealing with children in different stages of development.

Ten *characteristic parameters* showing the behaviour of the children were evaluated by means of a three-stage scale to assess the psychic symptoms and the therapeutic benefit derived from the drug under study. Normal behaviour was represented by the medium stage. The trial was open and took 4 weeks. Dosage was at first uniform (5 mg/kg body weight) and subsequently adjusted to the needs of the patient. No other drug was used. Regular interviews with the parents were made as supplementary therapeutic measures. Treatment was withdrawn in 1 case only.

See table I for the initial symptoms observed in the three groups of subjects.

Table I. Occurrence rates of symptoms (n = 27)

Anxious, dysphoric and depressed children with intentional inhibition	
Anxiety, fear	85%
Depression	89%
Dysphoria	92%
Isolated and contactless children in an outsider position	
Overadjustment	54%
Uncertainty, rigidity	69%
Isolation, lack of contact	81%
Children with lack of drive and concentration, exhibiting learning difficulties and low capability	
Poor performance	81%
Reluctance	46%
Lack of concentration	73%
Lack of drive	85%
Anxiety, dysphoria, depression	89%
Lack of contact, tendency to self-isolation	68%
Lack of drive and concentration	71%

Tabelle II. Symptoms prior to and following a 4-week treatment with sulpiride (n = 27)

Symptoms	Number of children	
	before starting treatment	at the end of treatment
Depression, sadness	23	6
Ill-humour, worry	24	11
Anxiety, fear	22	14
Isolation, lack of contact	21	9

Results

All psychic symptoms included in this study showed a statistically significant trend from pathologic values toward normal. Overall results were very good in 11%, good in 48% and poor in 41% of the cases.

In table II the response is classified according to the different psychic symptoms. After a 4-week treatment with sulpiride the following alterations from higher to lower degrees of severity were observed.

At the *beginning* of the treatment, the majority of the psychic symptoms were of a high degree of severity. In 6 cases they were assessed as severe, in 18 as moderately severe and in 3 as mild. In the first control examination (after 2 weeks) there was a clear trend toward lower degrees of severity. Upon *completion* of the trial one child had obtained freedom from psychopathological abnormalities, while mild disorders were seen in 14 cases and moderately severe disorders in 11 cases. Only 1 patient was still suffering from serious psychic problems.

Remission of the *psychosomatic symptoms* was already discernible in the first control examination. However, these alterations were not examined for statistical significance. Improvement was only expressed numerically.

Very good or good remissions were achieved in the following indications: mutism (decreased in 5 out of 6 cases); sleep disturbances (improved in 11 out of 13 cases); nocturnal anxiety (relieved in all 5 cases); paroxysms of weeping (reduced in 4 out of 6 cases); headache (improved in 3 out of 4 cases).

The following symptoms showed an unsatisfactory response or *none at all*, or the number of cases was too small for assessment: thumb-sucking, nail-biting, daintiness, enuresis, encopresis and genital manipulation.

After 2 weeks of treatment, the *emotional and verbal communication* of one-third of the children and adolescents had been clearly improved, so that psychotherapeutical measures could be considered. The dosage of 3–5 mg/kg body weight was found to be adequate for children over 6 and adolescents. Up to now, we have not had any experience in the treatment of infants. The syrup was above all preferred by the young school children.

The drug was *tolerated* very well in 21 cases and well tolerated by 6 patients. *Side effects* (dry mouth) were seen in one child only.

Discussion

The results of the present study show once more that sulpiride is a drug with a wide spectrum of effectiveness including the *actions of both antidepressants and major tranquilizers.* Regarding the use of psychotropics in young people with psychic problems, its major advantage is its very good or *good tolerance.* Side effects were rarely seen and usually minor. This is of considerable importance, not only with respect to the principle of 'nihil nocere', but also for the relationship between the physician and the parents, which may be greatly disturbed by undesirable side effects of a drug.

Summary

In the treatment of children with different psychopathological symptoms, a very good *response* was seen in 11% of 27 cases, while 48% had a satisfactory or poor response. The psychic symptoms showed a clearer trend to remission than the psychosomatic attributes.

The following *indications* were found to be most responsive to the treatment with sulpiride: depression, dysphoria, lack of contact, tendency to self-isolations, lack of drive and concentration; further, the symptoms mutism, disturbances of sleeping-waking rhythm including nocturnal anxiety.

References

Abe, H.: Sulpiride in school phobia. Psychiatria clin. *8:* 95–98 (1975).

Desbuquois, G.; Vialatte, J.; Paupe, J.; Rossignol, C.: Effets antiémétiques et eutrophiques du sulpiride chez l'enfant. Sem. Hôp. Paris *45:* 1852–1855 (1969).

Duche, D.J.; Delagrange, G.: Essai du sulpiride en neuropsychiatrie infantile. Gaz. méd. Fr. *14:* (1970).

Freyhan, F.A.: Psychomotilität, extrapyramidale Syndrome und Wirkungsweisen neuroleptischer Therapien (Chlorpromazin, Reserpin, Prochlorperazin). Nervenarzt *28:* 504–509 (1957).

Gittelman-Klein, R.: Recent advances in child psychopharmacology (Human Sciences Press, New York 1975).

Kohler, C.; Margon, G.; Maritan, C.: Essai du sulpiride en pédopsychiatrie. Rev. Neuropsych. infant. *19:* 751–758 (1971).

Kovacs, M.; Beck, A.T.: An empirical-clinical approach toward a definition of childhood depression; in Schulerbrandt, Raskin, Depression in childhood. Diagnoses, treatment and conceptual models (Raven Press, New York 1977).

Lang, J.L.; Le Strat, M.: Evaluation des effets cliniques du sulpiride en externat médico-pédagogique et hôpital de jour pour enfants. Rev. Neuropsych. infant. *19:* 759–772 (1971).

Lonchamp, D.; Raffi, A.; Noir, A.; Shirrer, J.; Colin, P.: La pathologie comportementale de l'enfant et son traitement par le sulpiride. Chron. méd., Lyon *47:* 4309–4312 (1971).

Mendlewicz, J.; Praag, H.M. van: Childhood psychopharmacology. Current concepts (Karger, Basel 1978).

Miller, B.; Wallis, H.: Über die Wirkungsweise von Sulpirid bei autistischen Kindern. Münch. med. Wschr. *119:* 667–669 (1979).

Nissen, G.: Depressive Syndrome im Kindes- und Jugendalter (Springer, Berlin 1971).

Nissen, G.: Psychopharmaka bei Verhaltensstörungen im Kindesalter; in Spiel, Die Therapie im Kindes- und Jugendalter (Egermann, Wien 1976).

Nissen, G.: Anxiety syndromes in children and their treatment with beta-blockade; in Kielholz, Beta-blockers and the central nervous system (Huber, Bern 1977).

Nissen, G.: The role of drugs in the treatment of behavioral disorders in children. (Huber, Bern 1978).

Picot, C.; Pincon, J.A.: Intérêt du sulpiride dans les troubles du comportement de l'enfant. Chron. méd., Lyon *46:* 3393–3396 (1970).

Puyelo, R.; Moreau, S.; Miedzyrzecki, J.: Expérimentation du sulpiride en pédopsychiatrie. Sem. Hôp. *46:* 93–96 (1970).

Rothenberger, A.: Zur Therapie kindlicher Verhaltensstörungen mit Sulpirid. Kinderarzt 7 (1976).

Werry, J.S.: Pediatric psychopharamcology (Brunner/Mazel, New York 1978).

Wiener, J.M.: Psychopharmacology in childhood and adolescence (Basic Books, New York 1977).

Prof. Dr. G. Nissen, Direktor der Klinik und Poliklinik für
Kinder- und Jugendpsychiatrie, Füchsleinstrasse 15, D–8700 Würzburg (FRG)

Adv. biol. Psychiat., vol. 7, pp. 173–186 (Karger, Basel 1981)

Second Generation of Antidepressants

B.E. Leonard

Pharmacology Department, University College, Galway, Ireland

Introduction

The chance discovery of imipramine nearly three decades ago heralded the introduction of pharmacopsychiatry and for the first time in human history, it was possible to use a drug to effectively treat depressive illness with a high degree of success. The introduction of imipramine also led to the game of 'molecular roulette' and the clinician eventually had at his disposal some twenty or so chemical analogues of imipramine with essentially the same degree of efficacy. The monoamine oxidase inhibitors clearly supplemented the tricyclic antidepressants when it was eventually found that they were most effective in the treatment of the atypical depression rather than the clinically well-defined reactive or endogenous depression. Thus, the world of psychiatric medicine was gradually perceiving the advantages and disadvantages of the tricyclic antidepressants and monoamine oxidase inhibitors when a new series of antidepressant drugs was introduced to the market. The cynic may say that as clinicians are only just beginning to learn how to properly use the first generation of antidepressants, it is premature to introduce the second generation of drugs about which we know even less! Such is the march of progress. Drug firms, like members of the medical profession, must make a living and just as it is no longer possible to intellectually challenge the view that the earth is round (although emotionally I find the idea of a flat earth more appealing), so we must face the fact that the tricyclic antidepressants and monoamine oxidase inhibitors have many deficits both in terms of their restricted efficacy and side effects. It is against this background that I wish to address myself to the following questions: (a) What are the limitations of the first generation of antidepressants and can we expect the second generation of these drugs to offer any real advantage? (b) By knowing the possible mechanisms of action of the various types of antidepressant drug currently available, do we have a better insight into the aetiology of depression? Clearly if we have some idea of the cause of depressive illness, it may be possible to develop antidepressant drugs in a more rational way than we do at present.

First- and Second-Generation Antidepressants: Their Advantages and Disadvantages

The widely discussed review of the efficacy of tricyclic antidepressants, monoamine oxidase inhibitors and electroconvulsive shock therapy (ECT) conducted by the Medical Research Council [1975] came to the conclusion that ECT was the most effective method of treating depression, closely followed by tricyclic antidepressants. By contrast, monoamine oxidase inhibitors were only slightly more effective than placebo. It is generally accepted that about 25% of severely depressed patients are resistant to conventional therapy and therefore, on this ground alone, it may be argued that drugs other than the tricyclic antidepressants are needed if only because current therapy is not optimally effective. The question remains as to how much a new antidepressant should be selected and in this respect one is reminded of the caution prompted by the editorial in the *British Medical Journal* [1978] which stressed that novel chemical structure or biochemical profile is not necessarily relevant to the choice of an antidepressant drug. To state the obvious, only man is known to suffer from depression and even though sophisticated studies of the behaviour of socially deprived primates, for example, may ultimately help in our understanding of psychiatric depression [*McKinney and Bunney,* 1979], at present the ultimate test for a new antidepressant is its therapeutic action in man. Because of this situation the search for new antidepressants has been dependent on the pharmacological, behavioural and biochemical profiling of drugs known to be therapeutically effective. It is not surprising, therefore, to find that most of the currently available antidepressants have a qualitatively similar pharmacological profile and therapeutic efficacy. In recent years, investigators have attempted to narrow this impasse by: (a) Quantitatively analysing the effects of new psychotropic drugs on the human EEG. Such profiling has enabled antidepressants such as mianserin to be discovered even though this drug was devoid of antidepressant action as assessed by conventional animal test [*Itil* et al., 1972]; (b) Studying the direct effect of psychotropic drugs on the metabolism of those biogenic amines (i.e. noradrenaline and serotonin) thought to be deranged in depression. Such an approach has led to the discovery of a range of new drugs with antidepressant activity, but which appear to show a specificity of action on either the noradrenergic or serotonergic system, e.g. zimeladine, nomifensin, trazodone, citalopram. (c) Looking for new animal models of depression which will enable drugs to be selected on the basis of their potential antidepressant activity in man rather than according to their specific mode of action on neurotransmitter synthesis (e.g. blockade of amine reuptake or enhancement of amine release). While it would be incautious to suggest that such a model has been developed, there is tentative experimental evidence that the rat olfactory bulbectomy model may partly fulfill this criterion [*Cairncross* et al., 1979; *Tuite and Leonard,* 1981]. This is not

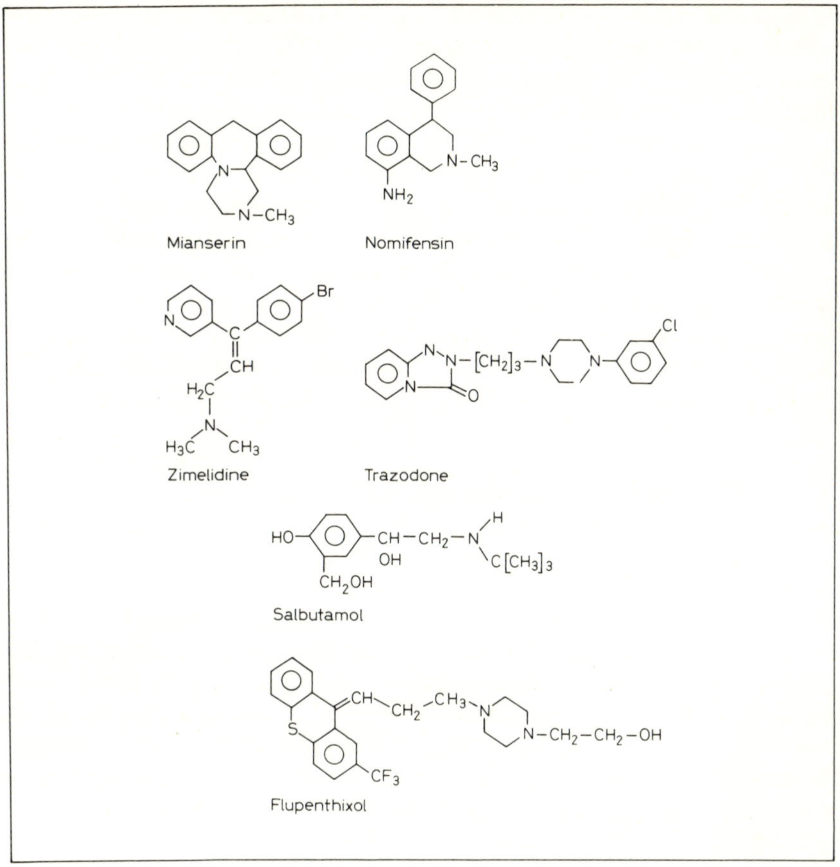

Fig. 1. Structures of some 'second-generation' antidepressants.

the place to describe details of the olfactory bulbectomy model which has been adequately covered in the references given above. Suffice it is to say that this model, unlike all other animal behavioural models described, has so far enabled all known antidepressants to be detected whereas psychotropic drugs which lack antidepressant activity are inactive in this test system. Furthermore, the activity of antidepressant is only detectable following their chronic administration, an effect which simulates the clinical situation whereby the drugs must be administered for a minimal period of 2 weeks before any clinical benefit is experienced by the patient.

The criteria which I have used to select some 'second-generation' antidepressants is necessarily rather arbitrary, but all the drugs subsequently detailed fulfill two or more of the following criteria: (a) Their antidepressant properties have been established within its last decade. (b) They show greater selectivity of action on either the serotonergic or noradrenergic system than the 'classical' dibenzazepine type of antidepressants (i.e. the 'tricyclics'). (c) They do not have a dibenzazepine type of chemical structure. (d) They have few anticholinergic side effects and are less cardiotoxic in man than the classical antidepressants such as amitriptyline. To illustrate the evolutionary development of antidepressant drugs, I will now outline the general pharmacological and biochemical properties of some recently introduced drugs that help to demonstrate their diversity of action and, as a consequence, thereby help to broaden our knowledge of the synaptic regulation of neurotransmitter metabolism. The drugs selected are mianserin, nomifensin, zimelidine, trazodone, salbutamol and flupenthixol (fig. 1).

Mianserin

The antidepressant properties of this drug, like these of imipramine and the monoamine oxidase inhibitor iproniazid, was discovered by serendipity rather than by the application of a hypothesis. Thus, initially mianserin was investigated for its peripheral antiserotonin and antihistamine properties [*van der Burg* et al., 1970] and was subsequently found to be a moderately potent serotonin and histamine antagonist [*Saxena* et al., 1971]. In the early clinical trials, this drug was found to have sedative properties and, following the initial studies of *Itil* et al. [1972], to produce a profile somewhat similar to impramine in the quantitative human EEF. Subsequent clinical studies have established the therapeutic efficacy of mianserin as an antidepressant in the treatment of both endogenous and reactive depression [*Coppen* et al., 1976]; some investigators have also suggested that the drug may have an anxiolytic [*Itil,* 1975] and anti-obsessional [*Vaisanen* et al., 1977] action. Details of the clinical profile of mianserin have been published elsewhere [*Peet and Turner,* 1978; *Brogan* et al., 1978]. The initial clinical studies of mianserin showed that the drug had fewer peripheral anticholinergic side effects than the conventional tricyclic antidepressants and also, unlike the latter group of drugs, did not affect the reuptake of noradrenaline into peripheral sympathetic storage sites [*Ghose* et al., 1976; *Coppen* et al., 1976]. These observations confirmed the pharmacological findings in animals that mianserin was inactive in the tremorine tests in vivo [*van Riezen,* 1972] and did not inhibit the reuptake of either serotonin or noradrenaline into the rat brain in vivo [*Leonard,* 1974]. With the exceptions of iprindol, mianserin was therefore the only established antidepressant that did not inhibit the reuptake of biogenic amines in vivo. Subsequent biochemical studies on mianserin showed that, after both acute and chronic administration, the drug increased the turnover of noradrenaline; all antidepressants that inhibit the reuptake of this amine decrease the turnover by a feedback inhibitory mechanism which operates from the post-synaptic receptor. Further studies clearly showed that, in vivo, mianserin increases the release of noradrenaline, the evidence being that the concentration of the principle noradrenaline metabolite normetamephrine was increased [*Fludder and Leonard,* 1979], as was 3-methoxy-4-hydroxyphenylglycol [*Sugrue,* 1980], following the adminstration of the drug. The mechanism whereby this facilitation of noradrenaline release is effected is uncertain but the most plausible explanation has been provided by the in vitro studies of *Baumann and Maitre* [1977] who showed that the drug increased the release of noradrenaline from preloaded synaptosomes; this has been confirmed recently by *Schoemaker* et al. [1980].

Despite the numerous studies which show that mianserin is a potent serotonin antagonist in the periphery [*Saxena* et al., 1971; *Vargaftig* et al., 1971; *van Riezen,* 1972] and in modifying central serotonin receptor activity [*Maj* et al., 1978] there is no evidence that it affects the uptake or turnover of this amine in vivo [*Leonard and Kafoe,* 1976], neither does it affect the rate of serotonin synthesis following a peripheral tryptophan load [*Earley and Leonard,* 1980]. It would appear therefore, that the evidence suggesting that the antidepressant action of mianserin is attributable to its ability to modify central serotonergic activity is inconclusive. The in vivo neuropharmacological evidence of *van Riezen* et al. [1976], showing that mianserin does not affect the PGO spike frequency changes in the cat brain following reserpine administration, also supports our biochemical studies that the drug does not interfere directly with serotonin receptor function in vivo. The only clinical biochemical data to suggest that mianserin may affect serotonin metabolism comes from the studies of *Coppen* et al. [1978] who showed that the uptake of labelled serotonin into the platelets obtained from depressed, untreated patients returned to control values following chronic drug therapy. As this effect was not observed in vitro, these investigators postulated that an active metabolite of mianserin was involved; there is evidence from the changes in noradrenaline turnover in the rat brain to support this view [*Kafoe* et al., 1976]. From this synopsis of the properties of mianserin, it is clear that the drug fulfills all four criteria which enable it to be classified a 'second-generation' antidepressant.

Nomifensin

Being an isoquinoline derivative, nomifensin is structurally unrelated to the dibenzazepine group of antidepressants. There is evidence from double-blind and placebo-controlled clinical trials that nomifensin is an effective antidepressant [*Matz,* 1980; *Bogie* et al., 1980; *Forrest* et al., 1977] while the minimal anticholinergic and sedative side effects, with no noticeable impairment of psychomotor performance, have led to the recommendation of the drug for the treatment of depression in ambulant outpatients and in the elderly [*Stonier and Wittels,* 1980].

The therapeutic profile of nomifensin has been attributed to its selective inhibitory effect on the reuptake of noradrenaline [*Schacht and Heptner,* 1974]. However, such specificity of action has been observed in vitro and while there is clear evidence that the drug also impedes noradrenaline reuptake in the rat brain in vivo [*Leonard,* 1980] there is also evidence that the 4-hydroxymetabolite of nomifensin significantly inhibits serotonin reuptake [*Kruse* et al., 1977]. Thus, following the long-term administration of this drug, one would anticipate that both noradrenaline and, to a lesser extent, serotonin metabolism would be affected. In addition to its effect on these two amines, nomifensin has been shown to increase the turnover of acetylcholine in vivo, a property which is not found with other antidepressants [*Robinson* et al., 1978]. It is not possible to state unequivocally that this effect on acetylcholine metabolism is unrelated to its antidepressant action although it has been postulated that it is due to a noradrenergic activation of cholinergic neurons [*Robinson* et al., 1978]. The contribution of the blockade of dopamine reuptake to the antidepressant action of nomifensin is controversial. *Hunt* et al. [1974] reported that nomifensin was twice as potent as benztropine in vitro and ten times as potent as *dl*-amphetamine in inhibiting the uptake of tritiated dopamine into synaptosomes from rat brain striata; *Samanin* et al. [1975] showed that nomifensin also blocked dopamine reuptake in vivo. More recent evidence from both animal [*Baumann and Maitre,* 1979] and human studies [*Scanlon* et al., 1977] suggests that nomifensin stimulates dopamine receptors which may contribute to facilitation of central dopaminergic activity. From such observations, the question arises as to whether the antidepressant effect of nomifensin is attributable to its action on the noradrenergic, dopaminergic or even the central serotonergic systems. It would seem reasonable to assume that the efficacy of nomifensin in the treatment of depressed patients showing severe psychomotor retardation is due to the ability of the drug to facilitate dopaminergic transmission although it must be emphasized that so far the drug has not been found to be effective in the treatment of severe Parkinsonism [*Bedard* et al., 1979]. Nevertheless, *Randrup and Braestrup* [1977] have shown that in vitro other antidepressants beside nomifensin are potent dopamine uptake inhibitors. From such studies, one would anticipate that dopamine agonists would have useful antidepressant properties while, conversely, dopamine receptor antagonists would lead to a further deterioration of the patient. As there is clinical evidence that the thioxanthine neuroleptic drug, fluphenthixol has useful antidepressant properties [*Younge* et al., 1976], it is difficult to postulate how both dopamine agonists and antagonists could act as antidepressants by a direct action on dopaminergic receptors. Claims have also been made that the benzamide neuroleptic *sulpuride* has antidepressant properties [*Niskanen* et al., 1975]. One possible explanation for the antidepressant effects of some neuroleptics could be in their ability to induce and a subsensitivity of the dopamine autoreceptors located on the soma and dendrites of dopaminergic neurons. *Serra* et al. [1979] proposed that tricyclic antidepressants might exert their therapeutic effects, at least in part, by acting in this way and this view has been supported by the investigation of *Chiodo and Antelman* [1980] who showed that only after chronic administration of amitriptyline and imipramine could the induction of dopamine receptor subsensitivity be observed. From such studies, it may be speculated that neuroleptics, such as flupenthixol, or antidepressants, such as nomifensin, could produce dopamine autoreceptors subsensitivity either by blocking these receptors (flupenthixol) or by stimulating them (nomifensin) thereby producing receptor desensitization. The net effect would be a reduction in dopaminergic transmission. Unless it is postulated that such an effect would have a disinhibitory action on noradrenergic and/or serotonergic transmission in those limbic regions of the brain that are malfunctioning in depression, it is difficult to link such a dopaminergic hypothesis of depression into the conventional 'two amine' hypothesis which implicates an abnormality of noradrenaline and/or serotonin metabolism of the disease.

In addition to its effects on at least four neurotransmitter systems in the mammalian brain, there is also evidence that this drug shares with dibenzazepine antidepressants an ability to potentiate the action of narcotic analgesics [*Gonzales* et al., 1980]. While 'classical' antidepressants have been used clinically to alleviate severe pain [*Dudley-Hart,* 1976; *Hughes* et al., 1963] for over a decade, it is only now apparent that the potentiation of narcotic analgesia, or the analgetic effect of the antidepressant itself, is attributable to the direct agonistic action of these drugs on central opiate receptors [*Biegon and Samuel,* 1980]; eight antidepressant drugs of diverse chemical structure and action on biogenic amine metabolism were found by the latter investigators to have a qualitatively similar effect on the opiate receptors with rat brain. How important such findings are to the therapeutic activity of antidepressants, including nomifensin, must await elucidation, but they do serve as a caution to those speculating that the mode of action of any antidepressant is related to a specific effect on one neurotransmitter system.

Zimelidine

This is a bicyclic compound derived from pheniramine. Both zimelidine and its demethylated secondary amine metabolite norzimelidine, have been shown to be potent inhibitors of serotonin reuptake in vitro [*Ross and Renyi,* 1977]; these drugs are also weak inhibitors of noradrenaline reuptake, but have no effect on the dopaminergic system. Both in vitro and in vivo data show that zimelidine is a more selective inhibitor of serotonin reuptake than chlorimipramine [*Ross* et al., 1976] and has only a weak anticholinergic action in man [*Coppen* et al., 1979a, b]. One feature of particular importance regarding the structure of zimelidine relates to the isomeric carbon in the side chain; only the cis-isomer is active in blocking serotonin reuptake [*Abrahamsson* et al., 1976].

The clinical efficacy of zimelidine has been established in a double-blind clinical trial against amitriptyline [*Coppen* et al., 1979a] but in a subsequent study, this group of investigators could find no significant relationship between the inhibition of serotonin uptake into the platelets of depressed patients being treated with zimelidine and the therapeutic outcome [*Coppen* et al., 1979b], neither was the plasma concentration of this drug or its active metabolite significantly correlated with the therapeutic outcome. Previous studies also established that the inhibition of noradrenaline uptake into the peripheral sympathetic system by amitriptyline was not correlated with the clinical response [*Coppen and Ghose,* 1976]. Such findings suggest that either there is no causal relationship between the action of a drug on the amine uptake systems and the subsequent antidepressant response or that the pharmacokinetic characteristics of the antidepressant in the CNS compartment differs from those occuring in the periphery. It is already established that norzimelidine has a plasma half-life of approximately 15 h in normal volunteers which is approximately three times greater than the parent compound [*Brown* et al., 1980]. Presumably the antidepressant response attributed to zimelidine is therefore more likely to be related to its metabolite.

Trazodone

Trazodone was the first triazolopyridine derivative to be produced with some antidepressant properties; it is structurally unrelated to any of the known 'second-generation' antidepressants. Some 15 clinical reports of the antidepressant effects of this drug have appeared in the literature in the period up to 1974 and have been summarized by *Saarma* [1974]. Subsequent studies in the USA and UK have tended to confirm the earlier expectations. The potent sedative action of the drug may contribute to its anxiolytic activity. From the clinical literature it would appear that trazodone, like the other 'second-generation' antidepressants described in this review, has fewer anticholinergic side effects than the conventional tricyclics.

The biochemical profile of trazodone resembles that if zimelidine and chlorimipramine in that it shows some specificity of action on the central serotonergic system. There is evidence that the drug impedes serotonin reuptake [*Garattini* et al., 1976] while other investigators have shown that it may have a biphasic effect on serotonin metabolism with low doses acting as an antagonist while high doses have an agonistic effect [*Baran* et al., 1979; *Maj* et al., 1979]. It has been suggested that the activity of trazodone may be attributed to an active metabolite(s); m-chlorophenylpiperazine (CPP), and the structurally related trazodone derivative etoperidone, have been implicated as possibilities. However, it must be emphasized that there is no unequivocal evidence to suggest that either of these compounds are produced from trazodone in vivo despite the fact that etoperidone has a similar biphasic action on the serotonergic system to trazodone

[*Przegalinski and Lewandowska,* 1979]. It will be of interest to see if the neurochemical profile of trazodone is an anyway related to the anxiolytic and antidepressant properties that have been implicated by the preliminary clinical data.

Other Specific Serotonin Uptake Inhibitors. This structurally diverse group of compounds include citalopram, nitralopram, danitracen, femoxetine and fluvoxamine. All show considerable potency and specificity in inhibiting the reuptake of serotonin into rat brain synaptosomes. However, it still remains to be established whether these compounds are effective antidepressants and, if so, whether there is any relationship between their effect on serotonin metabolism and their clinical efficacy.

Salbutamol.

One of the most unexpected clinical observations on the treatment of depression to be made in recent years was the discovery that salbutamol, the β-adrenoceptor agonist drug conventionally used for the treatment of bronchial asthma, had antidepressant properties. The antidepressant activity of this drug was noticed in a pilot study of *Lecrubier* et al. [1977] and confirmed by *Simon* et al. [1978]. However, it is only recently that *Lecrubier* et al. [1980] undertook a controlled study in which salbutamol was compared with chlorimipramine in a group of mixed depressives. These investigators found that salbutamol was effective with chlorimipramine but that the onset of antidepressant action of salbutamol was more rapid (approximately 5 days). This confirmed the findings of the initial 'open' studies as anticipated. The most frequent side effects observed were tremor, tachycardia, anxiety and insomnia but it is reported that these symptoms diminished in intensity within 2–3 days. While there is a suggestion from animal studies that β-adrenoceptor agonists may have antidepressant activity [*Frances* et al., 1978], no detailed investigations of the effects of such drugs on brain neurotransmitter metabolism appear to have been made.

A Speculation on the Mode of Action of 'Second-Generation' Antidepressants

It seems reasonable to assume that depression arises as a consequence of a neurotransmitter malfunction in the limbic areas of the brain. The question arises as to which neurotransmitter(s) and how do antidepressants alleviate the symptoms by modifying the neurotransmitter systems? This topic has been the subject of more detailed discussion elsewhere [*Leonard,* 1980]; but it would appear that emphasis is now being laid not so much upon the pre-synaptic events governing the synthesis and release of the neurotransmitter as upon the changes in the activity of the pre- and post-synaptic receptors that govern neurotransmitter release. The observation by *Wood and Coppen* [1980] that the post-synaptic (α_1) adrenoceptors are hypersensitive, while the pre-synaptic (α_2) receptors are hyposensitive in depression help to validate the concept of an altered receptor sensitivity being responsible for the symptoms of depression. *Pandey* et al. [1979] has shown that the activity of β-adrenoceptors is subnormal in drug-free depressed patients.

It is difficult to reconcile these preliminary clinical findings with our knowledge of the regulation of sympathetic activity in the brain. There is experimental evidence to support the view that the α_2-receptor modifies the release of noradrenaline; stimulation of this receptor-reducing release while inhibition increases it [*Stärke* et·al., 1977]. It has been suggested that pre-synaptic β-adrenoceptors are antagonistic to the α_2-receptors which might explain the fact that salbutamol could facilitate noradrenergic transmission by increasing the release of this amine via an agonist action on the pre-synaptic β-receptors. Undoubtedly the lag period between the commencement of antidepressant medication and the clinical response is associated with the readaptation of the pre- and post-synaptic receptor activity to the presence of an increased concentration of neurotransmitter which has been brought about by the antidepressant drug inhibiting the amine reuptake system (e.g. nomifensin, zimelidine, 'classical' antidepressants), facilitating noradrenaline release (e.g. mianserin), sensitizing the noradrenergic receptors (e.g. iprindole, thyroxine) or stimulating pre-synaptic β-receptors thereby facilitating neurotransmitter release (e.g. salbutamol). It is interesting to find that all antidepressants so far investigated, including those such as trazodone and chlorimipramine which show some selectivity of action on the serotonergic system, induce a subsensitivity of the noradrenaline-sensitive adenylate cyclase system of the rat brain following their chronic administration [*Sulser,* 1978, 1979; *Vetulani* et al., 1976; *Wolfe* et al., 1978; *Clements-Jewry* et al., 1980]. It may therefore be speculated that the primary action of all antidepressants is to induce initially a subsensitivity of the α_2-receptors as a consequence of their action on amine uptake mechanisms, etc. Once this subsensitivity is established, an event requiring several days of drug treatment, the release of noradrenaline is increased [*Crews and Smith,* 1978; *McMillen* et al., 1980]. If it is assumed that many of the symptoms of depression are a consequence of the post-synaptic β-receptors becoming supersensitive in the non-drug treated patients [*Sulser,* 1979] then it would be anticipated that the enhanced release of the neurotransmitter would lead to a desensitization of the post-synaptic receptors and thereby reduce the symptoms of the disease.

The recent emphasis on receptor readaptation mechanisms that show a time course which parallels the period necessary for the onset of antidepressant action, may ultimately lead to a more rational approach to the selection of antidepressant drugs. Use of the olfactory bulbectomy rat model may enable compounds with potential antidepressant activity to be selected for clinical study but at present it is not possible to predict the potential therapeutic potency of a drug from animal studies. However, it would appear that irrespective of the structure or supposed mode of action of the drug, most antidepressants are therapeutically of equal efficacy. The major clinical advantage of the 'second-generation' antidepressants appears to lie in their lack of cardiotoxicity and reduced anticholinergic side effects rather than in their improved efficacy in treating depression. Indeed, the major advance in the pharmacology of antidepressant medication will only arise when drugs are discovered that are more effective than electroconvulsive shock therapy. May be we shall have to await the third of fourth generation of antidepressants to fulfill such a role!

References

Abrahamsson, S.; Carlsson, A.; Carnmalm, B.; Dahten, B.: An X-ray study of the potentially antidepressant isomer of 3-(4-bromo-phenyl)-N,N-dimethyl-3(3-pyridyl) allylamine. Acta chem. scand. *30:* 609–612 (1976).

Baran, L.; Maj, J.; Rogoz, Z.; Skuza, G.: On the antiserotonin action of trazodone. Pol. J. Pharmac. *31:* 25–33 (1979).

Baumann, P.A.; Maitre, L.: Blockade of pre-synaptic α-receptors and of amine uptake in the rat brain by the antidepressant mianserine. Arch. Pharmacol. *300:* 31–37 (1977).

Bedard, P.; Pakres, J.D.; Marsden, C.D.: Nomifensin in Parkinson's disease. Br. J. clin. Pharmacol. *4:* suppl. 2, pp. 197S–190S (1977).

Biegon, A.; Samuel, D.: Interaction of tricyclic antidepressants with opiate receptors. Biochem. Pharmac. *29:* 460–462 (1980).

Bogie, W.; Hewett, A.J.; Baird, C.M.: Multicentre clinical comparison of nomifensine, maprotiline and imipramine in depressed patients in general practise. R. Soc. Med. Int. Congr. Ser. *25:* 95–102 (1980).

British Medical Journal: Choosing an antidepressant. *1:* 28 (1978).

Brogan, R.N.; Heel, R.C.; Speight, J.M.; Avery, G.S.: Mianserin: a review of its pharmacological properties and therapeutic efficacy in depressive illness. Drugs *16:* 273–301 (1978).

Brown, D.; Scott, D.H.J.; Scott, D.B.; Meyer, M.; Westerbund, D.; Lundstrom, J.: Pharmacokinetics of zimelidine. Eur. J. Pharmacol. *17:* 111–116 (1980).

Burg, W.J. van der; Bonta, I.L.; Delobelle, J.; Ramon, C.; Vargaftig, B.: A novel type of substituted piperazine with high anti-serotonin potency. J. med. Chem. *1:* 35–39 (1970).

Cairncross, K.D.; Cox, B.; Forster, C.; Wren, A.F.: Olfactory projection systems, drugs and behaviour: a review. Psychoneuroendocrinology *4:* 253–272 (1979).

Chiodo, L.A.; Antelman, S.M.: Tricyclic antidepressants induce subsensitivity of pre-synaptic dopamine auto-receptors. Eur. J. Pharmacol. *64:* 203–205 (1980).

Clements-Jewry, S.; Robson, P.A.; Chidley, L.J.: Biochemical investigations into the mode of action of trazodone. Neuropharmacology *19:* 1165–1174 (1980).

Coppen, A.; Ghose, K.: Clinical and pharmacological effects of treatment with a new antidepressant. Drug Res. *26:* 1166–1170 (1976).

Coppen, A.; Gupta, R.; Montgomery, S.; Ghose, K.; Bailey, J.; Burns, B.; de Ridder, J.J.: Mianserin hydrochloride: a novel antidepressant. Br. J. Psychiat. *129:* 342–345 (1976).

Coppen, A.; Montgomery, S.; Ghose, K.; Rama Rao, V.A.; Bailey, J.; Christiansen, J.;

Mikkleson, P.L.; Praag, H.M. van; Poel, F. van de; Minsker, E.L.; Kozulja, V.G.; Matussek, N.; Kungkunz, G.; Jorgensen, A.: Amitriptyline plasma concentration and clinical effect: A WHO Collaborative Study. Lancet *1:* 63–66 (1978).

Coppen, A.; Ramarao, V.A.; Swade, C.; Wood, K.: Zimelidine: a therapeutic and pharmacokinetic study in depression. Psychopharmacology *63:* 199–202 (1979a).

Coppen, A.; Ramarao, V.A.; Swade, C.; Wood, K.: Effect of mianserin hydrochloride on pheripheral uptake mechanism for noradrenaline and 5-hydroxytryptamine in man. Br. J. clin. Pharmacol. *5:* 135–175 (1979b).

Crews, F.T.; Smith, C.B.: Pre-synaptic alpha-receptor subsensitivity after long-term antidepressant treatment. Science *202:* 322–324 (1978).

Dudley-Hart, F.: The use of psychotropic drugs in rheumatology. J. med. Res. *4:* suppl. 2, pp. 15–20 (1976).

Earley, C.J.; Leonard, B.E.: Lack of effect of chronically administered mianserin on therserotonergic system of the rat brain. IRCS Med. Sci *8:* 360–361 (1980).

Fludder, J.; Leonard, B.E.: Chronic effects of mianserin on noradrenaline metabolism in the rat brain: evidence for a pre-synaptic α-adrenolytic action in vivo. Psychopharamcology *64:* 329–332 (1979).

Forrest, A.; Hewett, A.; Nicholson, P.: Controlled randomized group comparison of nomifensine and imipramine in depressive illness. Br. J. clin. Pharmacol. *4:* suppl. 2, pp. 215S–222S (1977).

Frances, G.; Peuch, A.J.; Simon, P.: Profil psychopharmamcologique de l.isoprenaline et du salbutamol. J. Pharmacol. *9:* 25–34 (1978).

Garattini, S.; de Gaetano, R.; Saminin, R.; Bernasconi, S.; Roncaglioni, M.C.: Effects of trazodone on serotonin in the brain and platelets of the rat. Biochem. Pharmac. *25:* 13–16 (1976).

Ghose, K.; Coppen, A.: Noradrenaline, depressive illness and the action of amitriptyline. Psychopharmacology *54:* 57–60 (1977).

Gonzales, J.P.; Sewell, R.D.L.; Spencer, P.S.J.: Antidepressants and pain: a review of some human and animal studies. R. Soc. Med. Int. Congr. Symp. Ser. *25:* 57–69 (1980).

Hunt, P.; Kannengiesser, M.H.; Raynand, J.P.: Nomifensine: a new potent inhibitor of dopamine uptake into synaptosomes from rat brain corpus striatum. J. Pharm. Pharmac. *26:* 370–371 (1974).

Hughes, A.; Chauvergne, J.; Lissilour, J.; Lagarde, C.: L'imipramine, utilisee comme antalgique majeur en carcinologie: etude de 118 cas. Presse méd. *71:* 1073–1080 (1963).

Itil, J.M.: The GB94 story; in Vossenaar, Proc. Symp. on Depressive illness and experiences with a new antidepressant drug, GB94, pp. 9–21 (Excerpta Medica, Amsterdam 1975).

Itil, J.M.; Polvan, N.; Hsu, W.: Clinical and EEG effects of GB94, a tetracyclic antidepressant. Curr. ther. Res. *14:* 395–413 (1972).

Kafoe, W.F.; de Ridder, J.J.; Leonard, B.E.: The effect of a tetracyclic compound, Org GB94, on the turnover of biogenic amines in rat brain. Biochem. Pharmac. *25:* 2455–2460 (1976).

Kruse, H.; Hoffmann, I.; Gerhards, H.J.; Leven, M.; Schacht, U.: Pharmacological and biochemical studies with three metabolites of nomifensine. Psychopharmacology *51:* 117–121 (1977).

Lecrubier, Y.; Jouvent, R.; Puech, A.J.; Simon, P.; Widlocher, D.: Effect antidepresseur d'un stimulant beta-adrenergique. Nouv. Presse méd. *6:* 2786 (1977).

Lecrubier, Y.; Puech, A.J.; Jouvent, R.; Simon, P.; Widlocher, D.: A beta-adrenergic stimulant (salbutamol) versus clomipramine in depression: a controlled study. Br. J. Psychiat. *136:* 354–358 (1980).

Leonard, B.E.: Some effects of a new tetracyclic antidepressant compound, Org GB94, on the metabolism of monoamines in the rat brain. Psychopharmacology *36:* 221–236 (1974).

Leonard, B.E.: Drug treatment in depression: a critical evaluation of the current situation. Proc. Soc. Med. Int. Congr. Symp. Ser. *25:* 1–9 (1980).

Leonard, B.E.; Kafoe, W.F.: A comparison of the acute and chronic effects of four antidepressant drugs on the turnover of serotonin, dopamine and noradrenaline in the rat brian. Biochem. Pharmac. *25:* 1936–1942 (1976).

Maj, J.; Palider, W.; Rawlow, A.: Trazodone, a central serotonin antagonist and agonist. J. neural Transm. *44:* 237–248 (1979).

Maj, J.; Sowinska, H.; Baran, L.; Gancarczyk, L.; Rawlow, A.: The central antiserotonergic action of mianserin. Psychopharmacology *59:* 79–84 (1978).

Matz, R.S.: Clinical trials with nomifensine in the USA. R. Soc. Med. Int. Congr. Symp. Ser. *25:* 73–80 (1980).

McKinney, W.T.; Bunney, W.E.: Animal model of depression. Archs gen. Psychiat. *21:* 240–248 (1979).

McMillen, B.A.; Warnack, W.; German, D.C.; Shore, P.A.: Effects of chronic desipramine treatment on rat brain noradrenergic responses to α-adrenergic drugs. Eur. J. Pharmacol. *61:* 239–246 (1980).

Medical Research Council Report by Clinical Psychiatry Committee: Clinical trial in the treatment of depressive illness. Br. med. J. *i:* 881–885 (1975).

Niskanen, P.; Tamminen, T.; Viukari, M.: Sulpiride vs. amitriptyline in the treatment of depression. Curr. ther. Res. *17:* 281–284 (1975).

Pandey; G.N.; Dysken, M.W.; Garter, D.L.; Davis, J.M.: Beta adrenergic receptor function in affective illness. Am. J. Psychiat. *136:* 675–678 (1979).

Peet, M.; Turner, P.: Proceedings of a symposium on mianserin, 1977.

Przegalinski, E.; Lewandowska, A.: The effect of etoperidone, a new potential antidepressant drug, on the central serotonin system. J. neural. Transm. *46:* 303–312 (1979).

Randrup, A.; Braestrup, C.: Uptake inhibition of biogenic amines by newer antidepressant drugs: relevence to the dopamine hypothesis of depression. Psychopharmacology *53:* 309–314 (1977).

Riezen, H. van; Different central effects of the 5-HT antagonists mianserin and cyproheptadine. Archs int. Pharmacodyn. Thér. *198:* 256–269 (1972).

Riezen, H. van; Proosdijk, J. van; Schonbaum, E.: Effects of various drugs supposed to interact with serotonin on PGO frequency changes induced by reserpine and 5-hydroxytryptophan; in Sicuteri, Schonbau, Monographs in neural sciences, vol. 3, pp. 37–44 (Karger, Basel 1976).

Robinson, S.E.; Cheney, D.L.; Costa, E.: Effect of nomifensine and other antidepressant drugs on acetylcholine turnover in various regions of rat brain. Arch. Pharmacol. *304:* 263–269 (1978).

Ross, S.B.; Ogren, S.O.; Renyi, A.L.: 2-Dimethylamino-1-(4-bromophenyl)-1-(3-pyridyl)propene (H10209), a new selective inhibitor of the neuronal 5-hydroxytryptamine uptake. Acta pharmac. tox. *34:* 152–166 (1976).

Ross, S.B.; Renyi, A.L.: Inhibition of neuronal uptake of 5-hydroxytryptamine and noradrenaline in rat brain by (Z) and (E)-3-(4-bromophenyl)-N,N-dimethyl-3-(3-pyridyl) allylamines and their secondary analogues. Neuropharmacology *16:* 57–63 (1977).

Saarma, J.: Trazodone – a review of literature. Mod. Probl. Pharmacopsychiat., vol. 9, pp. 95–109 (Karger, Basel 1974).

Samanin, R.; Bernasconi, S.; Garattini, S.: The effect of nomifensine on the depletion of brain serotonin and catecholamines induced respectively by fenfluramine and 6-hydroxydopamine in rats. Eur. J. Pharmacol. *34:* 377–380 (1975).

Saxena, P.R.; van Houwelingen, P.; Bouta, I.L.: The effects of mianserin hydrochloride on the vascular responses evoked by 5-hydroxytryptamine and related vasoactive sebstances. Eur. J. Pharmacol. *13:* 295–305 (1971).

Scanlon, M.F.; Gomez-Pan, A.; Mora, B.; Cook, D.B.; Dewar, J.H.; Hildyard, A.; Weightman, D.R.; Evered, D.C.; Hall, R.: Effects of nomifensine, an inhibitor of endogenous catecholamine re-uptake, in acromegaly, in hyperprolactinaemia and against stimulated prolactin release in man. Br. J. clin. Pharmacol. *4:* suppl. 2, pp. 191S–197S (1977).

Schacht, U.; Heptner, W.: Effect of nomifensine (Hoe 984). A new antidepressant, on uptake of noradrenaline and serotonin and on release of noradrenaline in rat brain synaptosomes. Biochem. Pharmac. *23:* 3413–3417 (1974).

Schoemaker, J.H.R.; Berendsen, H.H.G.; Stevens, H.J.T.; Nickolson, V.J.: Differences in pre-synaptic α-block and potential antidepressant activity between (+) and (–)-mianserin. Eur. J. Pharmacol. (in press, 1980).

Serra, G.; Argiolas, A.; Klimek, V.; Fadda, F.; Gessa, G.L.: Chronic treatment with antidepressants prevents the inhibitory effect of small doses of apomorphine on dopamine synthesis and motor activity. Life Sci. *25:* 415–420 (1979).

Simon, P.; Lecrubier, Y.; Jouvent, R.; Ruech, A.J.; Allilaire, J.F.; Widlocher, D.: Experimental. and clinical evidence of the antidepressant effect of a beta-adrenergic stimulant. Psychol. Med. *8:* 335–338 (1978).

Stärke, K.; Taube, H.D.; Borowski, E.: Pre-synaptic receptor systems in catecholaminergic transmission. Biochem. Pharmac. *26:* 259–268 (1977).

Stonier, P.; Wittels, P.Y.: A review of the cardiovascular effects of nomifensin. R. Soc. Med. Int. Congr. Symp. Ser. *25:* 18–22 (1980).

Sugrue, M.F.: Effects of acutely and chronically administered desipramine and mianserin on the clonidine-induced decrease in rat brain 3-methoxy-4-hydroxyphenylethylene glycol sulphate content. Br. J. Pharmacol. *69:* 299P–300P (1980).

Sulser, F.: Functional aspects of the norepinephrine receptor coupled adenylate cyclase system in the limbic forebrain and its modification by drugs which precipitate or alleviate depression: molecular approaches to an understanding of affective disorders. Pharmakopsychiatr. Neuro-Psychopharmakol. *11:* 43–52 (1978).

Sulser, F.: New perspectives on the mode of action of antidepressant drugs. Trends pharmac. Sci. *1:* 92–94 (1979).

Tuite, M. Leonard, B.E.: Anatomical, physiological and behavioural aspects of olfactory bulbectomy in the rat. Int. Rev. Neurobiol. (in press, 1981).

Vargaftig, B.B.; Coignet, J.L.; de Vos, C.J.; Grijsen, H.; Bonta, I.L.: Mianserin hydrochloride: peripheral and central effects in relation to antagonisms against 5-hydroxytryptamine and tryptamine. Eur. J. Pharmacol. *16:* 336–346 (1971).

Vetulani, J.; Stawarz, R.J.; Dingell, J.V.; Sulser, F.: A possible common mechanism of action of antidepressant treatments. Archs Pharmacol. *293:* 109–114 (1976).

Wolfe, B.B.; Harden, T.K.; Sporn, J.R.; Molinoff, P.B.: Pre-synaptic modulation of beta-adrenergic receptors in rat cerebral cortex after treatment with antidepressants. J. Pharmac. exp. Ther. *207:* 444–457 (1978).

Word, K.; Coppen, A.: Data presented at Spring Meeting of British Association of Psychopharmacology, London 1980.

Younge, J.P.R.; Hughes, W.C.; Lader, M.H.: A controlled comparison of flupenthixol and amitriptyline in depressed out patients Br. med. J. *i:* 1116–1118 (1976).

B.E. Leonard, DSc, Pharmacology Department, University College, Galway (Ireland)

Adv. biol. Psychiat., vol. 7, pp. 187–197 (Karger, Basel 1981)

A Trial of Real versus Simulated Electroconvulsive Therapy in 70 Depressed Patients

J.F.W. Deakin, T.J. Crow, E.C. Johnstone, C.D. Frith, P. Lawler, J.F. Macmillan, K. McPhersen, D.G.C. Owens, J.S. Price, M. Stevens

Clinical Research Centre, Division of Psychiatry, Harrow, England

Electrical induction of convulsions (ECT) in the treatment of psychiatric illness was introduced over 40 years ago by *Cerletti and Bini.* From this beginning until the present day the treatment has aroused considerable controversy. Following the decisive results of two large multicentre trials, few would argue that ECT is not considerably more effective than placebo tablets in the treatment of depression [*Greenblatt* et al., 1962; Medical Research Council, MRC, 1965]. However, the question of whether the convulsion is necessary for the antidepressant effects of ECT and possible long-term cognitive impairments remain controversial issues. At Northwick Park Hospital a double-blind controlled trial of real versus simulated ECT in depression was instigated to clarify these issues.

Five studies in the earlier literature and two recent studies have attempted to assess the necessity of the convulsion in the clinical efficacy of ECT; four report real ECT superior to simulated and three found no differences.

Cronholm and Ottoson [1960] compared the antidepressant effects of ECT (46 patients) with ECT in which the convulsion was shortened with intravenous lidocaine (23 patients). Lidocaine apparently attenuated the antidepressant response, but a major defect of the trial is that patients were allocated alternately to the groups and a third group was added some 2 years into the trial and, to increase the numbers in the group, successive patients were allocated to it. The trial, although admirable in many respects, was thus neither blind nor randomized. *Ulett* et al. [1956] compared photoshock and electroshock treatments with a control procedure involving sedation and found the convulsive treatments superior. However, the patient sample included schizophrenic as well as depressive illness. *Robin and Harris* [1962] found real ECT to be markedly superior to simulated ECT in 30 depressed patients. In contrast to these studies, *Brill* et al. [1959] and *Miller* et al. [1953] found no difference between simulated and real ECT in a heterogeneous population of patients.

Table I. Design features of the Northwick Park trial

Number of patients completing treatment	62
Diagnosis	Endogenous depression by PSE, *Feighner* et al. [1972], and MRC [1965] criteria
Placebo procedure	8 times anaesthesia and muscle relaxant
Active treatment	8 times electrically induced convulsions under anaesthesia and muscle relaxation
Allocation of treatments	random within clinical subgroups
Patients 'blind' to treatment?	all anaesthetized
Raters 'blind' to treatments?	different doctors rated and treated patients
Antidepressant medication	none during the treatment phase

Two recent studies have failed to clarify the issue of the necessity of the convulsion. *Freeman* et al. [1978] randomly allocated 40 depressed patients to two simulated or two real ECT followed, in both groups, by real ECT treatments whose number was determined according to the patients' conditions by an independent psychiatrist. More ECT treatments were judged necessary for the patients which had received two initial simulated ECT than those which received real ECT throughout. However, the basis on which these judgements were made is not clear for Hamilton depression ratings, differed only after the initial two treatments and converged thereafter. The results were interpreted as indicating the convulsion is the critical component of ECT. However, this conclusions seems weakened by the fact that the 4 patients who dropped out of the trial, presumably because of an unsatisfactory response, were all in the real ECT group. This raises doubts as to the validity of using numbers of subsequent ECTs as the outcome measure.

In contrast to the results of *Freeman* et al. [1978], *Lambourne and Gill* [1978] found no differences between simulated and real ECT in depressed patients. Although a weaker form of stimulus was used than in the study by *Freeman* et al. [1978], *Lambourne and Gill* [1978] assert that convulsions were induced in every real ECT treatment. It is worth nothing that the Hamilton scores of all groups of patients in these two recent trials are closely similar throughout [*Crow and Johnstone,* 1980] despite the presence of the group in the study by *Lambourne and Gill* [1978] which did not receive any convulsions.

In view of the indecisive and contradictory results in the literature, a trial of simulated versus real ECT in depressed patients was carried out at Northwick Park Hospital with the aim of elucidating the role of the convulsion in the antidepressant efficacy of ECT. A design (table I) was used which attempted to overcome the uncertainties of interpretation that have arisen in previous attempts.

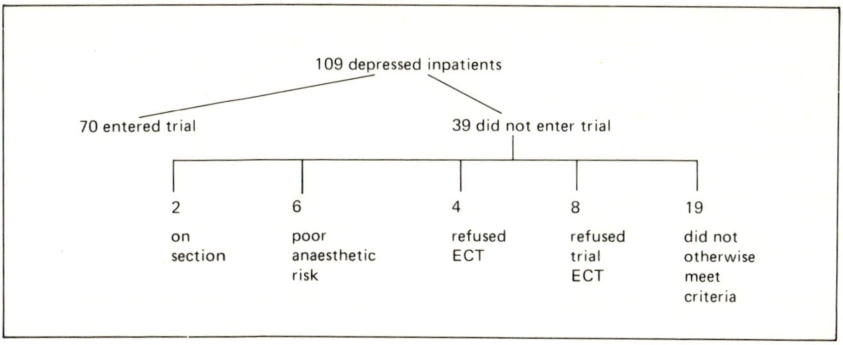

Fig. 1. Reasons for the 39 patients not to enter the trial.

The design of the trial afforded a good opportunity to investigate the question of whether convulsions induced undesirable long-term side effects on cognitive functions – some reports indicate this may be the case [*Halliday* et al., 1968; *Teuber* et al., 1976] whereas others have failed to find objective longterm impairments [*Squire and Chace,* 1975]. A battery of psychological tests were, therefore, administered before, during, and 6 months after the treatments.

Methods

Patients

A total of 109 patients referred to the unit for an acute episode of depression were considered for inclusion in the trial. 70 of these patients entered the trial and 39 did not for the reasons indicated in figure 1. The criteria for inclusion in the trial are as follows: The patients had to conform to the criteria used in the MRC trial [1965], which demonstrated the antidepressant efficacy of ECT, and to be likely to respond to ECT on the basis of the Newcastle scale for predicting a response to ECT. Diagnostic criteria were the present-state examination, the criteria for primary affective disorders by *Feighner* et al. [1972] and the Newcastle [*Carney* et al., 1965] criteria for endogenous depression. Some characteristics of the patients thus selected are shown in table II.

Treatments and Allocation

Real ECT was administered under methohexitone anaesthesia (1.5 mg/kg) and muscle relaxation (suxamethonium, 0.5 mg/kg). Convulsions were induced using an Ectron mark 4 machine set at waveform 1. This delivers 195 V for 1.7 s with a 'chopped' sinusoidal biphasic waveform. A bifrontal electrode placement was used. Induction of a convulsion was confirmed on every occasion by inflating a sphygmometer cuff to 200 mm Hg just before administration of suxamethonium. The convulsion was thus observed unmodified distal to the cuff in one forearm. Simulated ECT was identical to real ECT except that no current was passed. Treatments were administered on Mondays and Thursdays until a total of eight treatments per patients had been given. The psychiatrist administering ECT had no other contact with patients.

Table II. Some characteristics of the selected patients

Females	52
Males	18
Mean age	49.4 years
Previous depressions	46
Previous mania	7
Previous ECT (for earlier episode)	15

Table III. Cells used for substratification of patient sample prior to randomization.

	Deluded	Not deluded
Agitated	8 (6)	12
Retarded	8	17 (13)
Neither or both	10 (8)	15

Numbers in each cell show the numbers starting, those in parentheses the numbers finishing if different.

Table IV. Memory tests

Test	Timing of the test
Subjective memory	before ECT, after 8 ECT, at 6 months
Subjective concentration	before ECT, after 8 ECT, at 6 months
Vigilance	before ECT, after 4 ECT, after 8 ECT
Remote memory	before ECT, after 5 ECT, at 6 months
Semantic memory	1, 2, 4 and 7 h after ECT
Memory for ward lists	before ECT, after 8 ECT, at 6 months
Non-verbal	after 4 ECT
Learning labels for faces	5 trials/day before 5th ECT, 1 trial 2 h after 5th ECT, and 5 trials at 6 months

Table V. Reasons for non-completion of trial

	Real ECT	Simulated ECT
Failure to progress	1	3
Withdraw consent	1	1
Became manic	1	–
Refinal vascular accident	1	–

The patients were allocated to 1 of 6 clinical subgroups according to the presence or absence of delusions, the presence of agitation, retardation or both or neither of these (table III). Within each clinical subgroup the allocation real or simulated ECT was random. This ensured clinically comparable patients in the real and simulated ECT groups.

Assessments

Hamilton depression ratings [*Hamilton,* 1960] were carried out before treatment at weekly intervals throughout treatment and at 1 and 6 months after completion of treatment. Ratings were carried out by psychiatrists who were unaware of and unconnected with treatment 24 h after the preceding treatment (i.e., on Tuesdays and Fridays). Leeds self-rating scales were completed by patients at the same times as the Hamilton ratings.

An independent psychiatrist assessed whether patients were mentally fit to persist with the trial after four treatments.

No antidepressant medication was administered during the treatment phase. Nitrazepam was administered to all patients at night and diazepam 10 mg as required. After the completion of eight treatments the patients were treated with whatever antidepressant medication the patients' conditions warranted, including further ECT and tricyclics.

Psychological Tests

Tests and time of administration are listed in table IV.

Results

Effects of the Convulsion during the Treatment Phase

70 patients entered the trial and 62 completed the eight treatments. Reasons for non-completion are given in table V. Hamilton depression ratings for real and simulated ECT groups are shown in figure 2. These ratings were subjected to a two-way analysis of variance (treatment, time) for repeated measures on one factor (time; 5 Hamilton scores per patient). The main effect of time is highly significant ($F = 125.3$; 4, 240; $p < 0.001$) indicating significant improvement over the treatment period in both groups. There was no significant main effect of treatment since the scores of the two groups do not begin to diverge until after week 1. However, a significant interaction between treatment and time was found ($F = 4.07$; 4, 240; $p < 0.01$) due to the more rapid rate of improvement of the real ECT group compared to the simulated group.

It is clear, therefore, that the convulsion had an overall effect during the course of treatment in accelerating rate of decline of Hamilton depression scores. However, the effect was not large in relation to the improvements shown by both groups during the treatment phase.

Is the Effect of the Convulsion Sustained?

Hamilton depression ratings at the completion of treatment, at 1 and 6 months after completion of treatment are shown in figure 3. It is clear that the overall superiority of real compared to simulated ECT at the completion of treatment has disappeared 1 month later and is not apparent at 6 months.

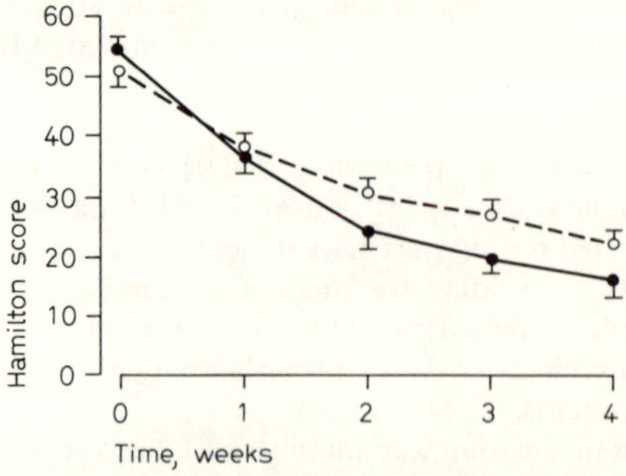

Fig. 2. Hamilton scores of real (•) and simulated (○) ECT groups during the treatment phase.

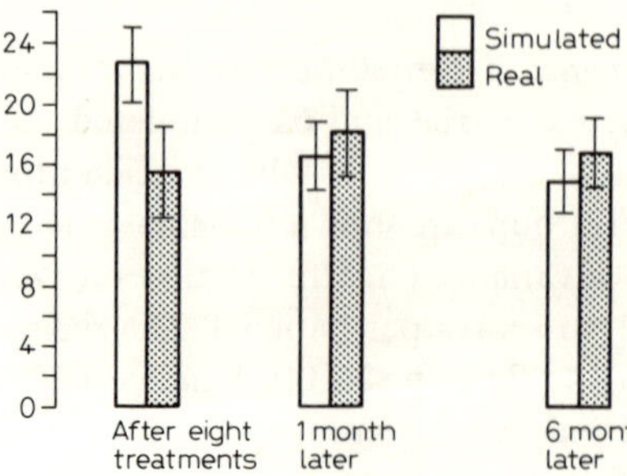

Fig. 3. Hamilton depression ratings at completion of treatment and 1 and 6 months later. N = 62 at completion; n = 57 for 1- and 6-month ratings.

Is the Convulsion More Effective in Some Patients than in Others?

To investigate this question, improvement scores from pretreatment ratings to those at completion and at 1 and 6 months after completion of treatment were subjected to analysis of variance with separate factors for treatment, presence or absence of delusions, and agitation/retardation. The statistical interactions between these clinical factors and treatment group are shown in table VI. It can be seen that at all three times there is a significant interaction. This indicates that the effect of the convulsion is significantly different in the different

Table VI. Interaction of treatment with presence of initial delusions, retardation, agitation

Timing of assessment	Direction of effect			F value	Degrees of freedom	Significance
	deluded	retarded	agitated			
At the end of eight treatments	+	+	–	3.7	6.54	< 0.01
After 1 additional month	+	–	–	4.3	6.54	< 0.01
After 6 additional months	+	+	–	4.1	6.49	< 0.01

+ = ECT > simulated; – = ECT < simulated.

Table VII. Treatment in months after trial

	Real	Simulated
Tricyclics	16	13
Other antidepressants	1	4
Anxiolytics	9	7
Lithium	1	2
ECT	1	2
Nil	3	2

Table VIII. Results of memory tests

Subjective memory	NS
Subjective concentration	NS
Vigilance[1]	slight ↑ errors for real ECT $p < 0.10$
Remote memory	NS
Semantic memory[1]	NS
Memory for word lists	NS
Non-verbal[1]	real ECT much worse
Learning labels for faces	ECT group impaired during the course: at 6 months NS

[1] Done during the course only.

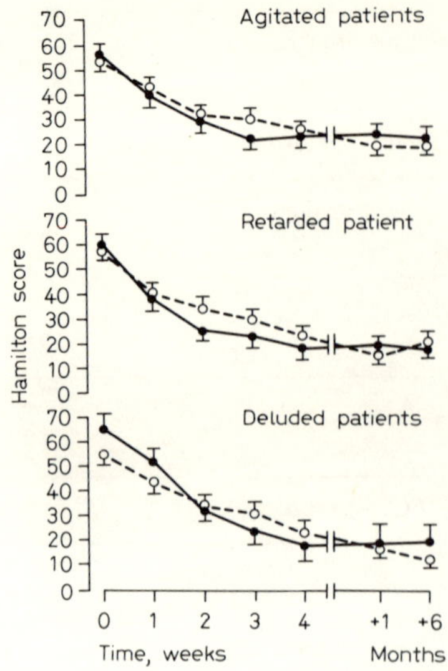

Fig. 4. Hamilton depression ratings in agitated, retarded, and deluded subgroups. ● = Real ECT; ○ = simulated ECT.

subgroups (fig. 4). At the completion of treatment the effect of the convulsion is greatest in the deluded subgroup and indeed deluded patients receiving the convulsion have significantly greater improvement scores than deluded patients receiving simulated ECT ($p < 0.02$, t test). Despite the significant interactions at 1 and 6 months, there are no significant differences between real and simulated ECT improvement scores within the deluded, agitated, or retarded subgroups. The significant differential effects of treatment on improvement scores across clinical subgroups is due to the fact that at all three times deluded patients tend to respond *relatively* more favourably to the convulsion than do agitated patients, who respond unfavourably to real ECT, and retarded patients, who show positive and negative responses to the convulsion depending on the time of rating. These patterns of response to treatment give rise to the significant interaction terms in table VI.

Does the Convulsion Affect Subsequent Treatment?

Table VII shows the treatments prescribed in the month following completion of real or simulated ECT. Treatment at this stage and thereafter was unrestricted by the design of the trial. It is apparent from table VII that the convulsion did not modify subsequent treatment.

Does Convulsion Affect Cognition?

The results of the psychological tests are shown in table VIII and will not be dealt with in detail. Of most interest are the results of a test of memory in which subjects have to associate names, occupations, and geographical areas with photographs of faces (e.g., 'Brenda is a barmaid from Oxford'). During the treatment phase the patients on real ECT were significantly impaired in learning and recalling this task. However, when the test was repeated with another set of faces 6 months later, no group differences emerged. It is also of some interest that tests of remote memory failed to reveal group differences at any time.

Discussion

The results show that the convulsion is a component of electroconvulsive therapy with significant antidepressant activity. The question of whether other components of ECT (general anaesthetic, muscle relaxant) contribute to the efficacy of the procedure, by pharmacological or psychological means, cannot be answered because an untreated group was not included in the design.

While the results show some similarities with those of the study by *Freeman* et al. [1978], there are important differences. *Freeman* et al. [1978] report significant differences between real and simulated ECT after two treatments, whereas no difference was detectable at this time (week 1 rating, fig. 2) in the present study. Furthermore, in the study of *Freeman* et al. [1978] clinicians, at the end of the treatment phase, were apparently able to detect a group difference (although there were no differences in Hamilton ratings) and administered more ECTs to the group which had received two simulated ECTs. In contrast, the group of patients which received no convulsions in the present study did not subsequently require more antidepressant therapy than the real ECT group (table VII). This would seem commensurate with the small size of the group difference in relation to the large improvement shown by both groups over the treatment period.

Lambourne and Gill [1978] found no significant differences between real and simulated ECT in depressed patients, in contrast to the present study. However, in their smaller patient sample the trend was in favour of the real ECT group. In addition, their patients included some outpatients and some who did not fulfill the Carney prognostic index for a favourable response to ECT. On the other hand, initial Hamilton scores were similar to those in the present study.

In contrast to both recent studies, the patients in this study were subdivided according to the clinical characteristics of the depression and it is clear that much of the significance of the present result is due to the particularly favourable response to convulsions in patients with delusions. Indeed, there was no difference between real and simulated ECT in patients without delusions. Thus, the large proportion of patients with endogenous depression, who do not have delusions, may not benefit from the convulsions in ECT.

In addition to the small magnitude and selective nature of the benefical effects of the convulsion, these effects were short-lived and had permanently disappeared 1 month after completion of treatment and had no influence on subsequent medication. It is, therefore, difficult to attribute the long-term benefical effect of ECT in reducing morbidity and mortality [*Avery and Winokur,* 1976] to the convulsion.

Patients not infrequently report long-lasting memory impairments following ECT [*Squire and Chace,* 1975] and forceful claims for such deleterious effects of ECT have been made [*Friedberg,* 1977]. No objective evidence for this point of view was obtained in the present study. For example, performance of the memory for faces test was unimpaired 6 months after ECT, although, in common with many other tests, performance was impaired during the course of ECT. Furthermore, no impairments on a test of remote memory were detected although ECT treatment patients complain of weakness in their remote personal memory. Since memory tests were carried out in patients for whom the trial was their only experience of ECT, it remains a possibility that multiple courses of ECT may cause long-term cognitive impairments.

Conclusions

A trial of simulated versus real ECT was carried out in patients suffering from endogenous depression according to three sets of clinical criteria. 31 patients received a course of eight electrically induced seizures under general anaesthesia and muscle relaxation. 31 patients received identical treatments except that the convulsion was omitted (simulated ECT). No other antidepressant treatments were administered during the treatment phase. Patients treated with real ECT showed a greater decline in Hamilton depression ratings over the 1-month treatment period than those receiving simulated ECT ($p < 0.01$). The beneficial effects of the convulsion were largely confined to the subgroup of patients with depressive delusions. 1 and 6 months after completion of treatment of overall superiority of real compared to simulated ECT was no longer apparent, and the treatments did not influence subsequent medication or further ECT treatment. Psychological tests did not reveal any long-term impairments of memory as a result of the convulsions.

The results indicate that while a single course of electrically induced convulsions produces no lasting impairment of memory, its therapeutic effects are of smaller magnitude and less long lasting than has sometimes been thought.

References

Avery, D.; Winokur, G.: Mortality in depressed patients treated with electroconvulsive therapy and antidepressants. Archs gen. Psychiat. *33:* 1029–1037 (1976).

Brill, N.Q.; Crumpton, E.; Eiduson, S.; Grayson, H.M.; Hellman, L.I.; Richards, P.A.: Relative effectiveness of various components of electroconvulsive therapy. Archs Neurol. Psychiat. *81:* 627–635 (1959).

Carney, M.W.P.; Roth, M.; Garside, R.F.: The diagnosis of depressive syndromes and the prediction of ECT response. Br. J. Psychiat. *111:* 659–674 (1965).

Cronholm, B.; Ottoson, J.-O.: Experimental studies of the therapeutic action of electroconvulsive therapy in endogenous depression. Acta psychiat. scand. *35:* suppl. 145, pp. 69–101 (1960).

Crow, T.J.; Johnstone, E.C.: The case against the electricity; in, Palmer, Electroconvulsive therapy; an appraisal (Oxford University Press, London, in press

Feighner, J.P.; Robbins, E.; Guze, S.B.; Woodruff, R.A.; Winokur, G.; Munoz, R.: Diagnostic criteria for use in psychiatric research. Archs gen. Psychiat. *26:* 57–63 (1972).

Freeman, C.P.; Basson, J.V.; Crighton, A.: Double-blind controlled trial of electroconvulsive therapy (ECT) and simulated ECT in depressive illness. Lancet *i:* 738–740 (1978).

Friedberg, J.: Shock treatment, brain damage, and memory loss. A neurological perspective. Am. J. Psychiat. *134:* 1010–1014 (1977).

Greenblatt, M.; Grosser, G.H.; Wechsler, H.: A comparative study of selected antidepressant medications and ECT. Am. J. Psychiat. *119:* 144–153 (1962).

Halliday, A.M.; Dawison, K.; Browne, M.W.; Kreger, L.C.: Comparison of the effects on depression and memory of bilateral ECT and unilateral ECT to the dominant and non-dominant hemispheres. Br. J. Psychiat. *114:* 997–1012 (1968).

Hamilton, M.: A rating scale for depression. J. Neurol. Neurosurg. Psychiat. *23:* 56–62 (1960).

Lambourne, J.; Gill, D.: A controlled comparison of simulated and real ECT. Br. J. Psychiat. *133:* 514–519 (1978).

Medical Research Council: Report by the Clinical Psychiatry Committee. Clinical trial of the treatment of depressive illness. Br. Med. J. *i:* 881–886 (1965).

Miller, D.H.; Clancy, J.; Cumming, E.: A comparison between unidirectional current non-convulsive electrical stimulation given with Reiter's machine, standard alternating current electroshock (Cerletti method), and pentothal in chronic schizophrenia. Am. J. Psychiat. *109:* 617–621 (1953).

Robin, A.A.; Harris, J.A.: A controlled comparison of imipramine and electroplexy. J. ment. Sci. *106:* 217–219 (1962).

Squire, L.R.; Chace, P.M.: Memory functions six to nine months after electroconvulsive therapy. Archs gen. Psychiat. *32:* 1557–1564 (1975).

Teuber, H.L.; Corkin, S.; Twitchwell, T.E.: A study of cingulotomy in man. A report to the National Commission for the Protection of Human Subjects of Biomedical and Behavioural Research (1976).

Ulett, G.A.; Smith, K.; Gleser, G.C.: Evaluation of convulsive and subconvulsive shock therapies utilizing a control group. Am. J. Psychiat. *112:* 795–802 (1956).

J.F.W. Deakin, MD, Clinical Research Centre, Division of Psychiatry, Wartford Road, Harrow, Middlesex HA1 3UJ (England)

Adv. biol. Psychiat., vol. 7, pp. 198–207 (Karger, Basel 1981)

Plasma Levels of Antidepressants and Therapeutic Effects[1]

Donald S. Robinson[a], *Thomas B. Cooper*[b], *Alexander Nies*[c], *Diantha B. Howard*[d], *John Corcella*[a]

[a]Marshall University School of Medicine, Huntington, W.V., USA; [b]Rockland Research Institute, Orangeburg, N.Y., USA; [c]Veterans Administration Medical Center, Coatesville, Pa., USA; [d]University of Vermont, Burlington, Vt., USA

There has been longstanding investigative interest in the clinical utility of tricyclic antidepressant (TCA) plasma levels since a sensitive assay for desipramine was introduced in 1967 by *Hammer and Brodie* [1]. Since that time, more specific and sensitive assays for TCAs using gas-liquid chromatography have permitted measurement of both tertiary and secondary amines of this drug class [2, 3]. The lack of uniform and predictable therapeutic response to TCAs has provided a major impetus to exploring the value of plasma assays as a means to improve drug response rate. Bettering therapeutic response by regulating dosage using blood levels would seem a reasonable strategy because of large interindividual differences in steady state plasma levels at standard TCA doses due to both genetic and environmental influences on drug metabolism.

There are approximately 40 published reports of studies of blood concentration of antidepressant drugs and clinical effects, as recently reviewed by *Amdisen* [4]. As this author points out, findings have been conflicting, with a trend for the larger sample size studies more recently reported to show little correlation of drug concentration and therapeutic response. Other recent reviews also have reached the conclusion that definitived relationships have not been established for any of the TCAs and have emphasized the need for appropriate additional studies to resolve the many discrepancies [5–7].

This failure to demonstrate a predictive value of plasma TCA determinations could be accounted for in part by such factors as imperfect experimental design of many reported clinical trials, differing and heterogeneous samples of depressed patients making comparisons and extrapolations difficult, and the large number of uncontrollable variables intrinsic to any antidepressant trial. Another important contributing factor could be the recent demonstration of other active

[1] Supported in part by Grant R01 MH27836 from the National Institute of Mental Health. Dr. *Robinson* is a Burroughs Wellcome Fund Scholar in Clinical Pharmacology.

metabolites of TCAs which are present in plasma, brain and other tissues in significant concentrations [8–10].

One of the more controversial drugs of this class in terms of the value of plasma determinations has been amitriptyline, a drug we have been studying for several years in the treatment of depressed outpatients. There have been at least 10 reported studies of amitriptyline (AMI) plasma levels in the treatment of depressive illness. Five of these have claimed to define a therapeutic range, or to find a significant correlation of plasma AMI and nortriptyline (NT) levels with improvement [11–15]. Except for the last, these have been small sample size studies. Five recent studies have uniformly failed to discern an association between drug level and clinical response to amitriptyline [16–20]. Of the latter group, two studies were also double-blind and controlled trials [19, 20]. and fixed dosages were employed in all. These five negative studies totalled 172 AMI-treated patients.

The somewhat disappointing findings of late have led some investigators to search for refinements of plasma level determinations in an attempt to detect meaningful relationships of clinical value. One strategy has been to assess the rate of demethylation of amitriptyline and relate it to clinical response. Because of the recent capability of measuring hydroxymetabolites of AMI [21] and imipramine [10] in plasma, another has been to investigate the possible contributions of other 'active' TCA compounds to overall pharmacologic and therapeutic effects in man.

We have been conducting an ongoing double-blind controlled clinical trial of amitriptyline vs. phenelzine in depressed outpatients [19]. In a previous report of this study, we failed to find significant overall correlations between plasma TCA concentrations and improvement. On dividing patients into subsamples as to those scoring more highly on endogenous features (Hamilton Rating Scale [HRS] ≥ 24, n = 25) and those patients scoring lower (HRS < 24, n = 22), there emerged a pattern of weak positive correlations of plasma AMI level and improvement in several major symptom scales in the high HRS group. We now report on a further examination of these data with regard to demethylation (NT/AMI) ratio and also additional analyses of 10-hydroxy-metabolites of AMI and NT in a recent patient sample.

Materials and Methods

The AMI-treated patients in this study ranged in age from 19 to 67 years and were referred largely from primary care physicians for psychiatric evaluation of symptoms of depression. Patients with depressive illnesses meeting Research Diagnostic Criteria for primary depressive disorder were included if the patient's symptomatology was sufficiently severe and/or chronic to warrant drug treatment. Patients with secondary depressions and short-lived depressive responses to life stresses were excluded. The initial Hamilton Depression scores (17-item) after a 1-week drug-free observation period averaged 24.1 ± 5.2, indicating that patients in this sample were moderately to severely symptomatic.

After a screening interview by one of the project psychiatrists and obtaining informed consent, patients underwent a 1-week drug-free washout period. When the patient was re-evaluated after 7 days, drug treatment by random, double-blind assignment was initiated if symptoms were persisting. A Standard Depression Interview (SDI) was administered to each patient by a trained research nurse. This observer-rated structured interview contains the Hamilton Depression Items [22] and additonal items occurring in depressive illness less typically endogenous [23]. Patients also completed the subjective self-rated Symptom Distress Checklist-90 (SCL-90) [24]. Treatment was begun with AMI 75 mg/day for 5 days followed by 150 mg/day for the remainder of a 6-week treatment period. Since this was part of double-blind comparative trial of amitriptyline and phenelzine, patients were treated double-blind with respect to both antidepressant drug assigment and plasma TCA levels.

The SCL-90 was completed at each 2-week clinic visit and the observer-rated SDI was repeated at the end of the 6-week drug treatment period. A side effect checklist and a global clinical rating were also completed by the treating psychiatrist at each patient visit.

Blood was collected in a glass free system using EDTA anticoagulant. AMI, 10-hydroxy-amitriptyline (10-OH-AMI), NT, and 10-hydroxy-nortriptyline (10-OH-NT) concentrations were determined in plasma obtained before and after 2, 4 and 6 weeks of drug treatment. No vacuum container tubes were used in any of the blood collections. Plasma was promptly separated from the blood by centrifugation and stored at –15 °C for analysis. AMI and NT concentrations were measured by gas-liquid chromatrography using a nitrogen detector by the method of *Cooper* et al. [3] and the 10-hydroxy metabolites of AMI and NT by a modification of the method of *Borga and Garle* [21]. This method has a coefficient of variation of a <5% at the 5 ng/ml level. All samples were analyzed in duplicate and the mean value reported.

Results

As previously reported, there was no evidence of significant correlations of AMI, NT, or AMI plus NT (total TCA) levels with improvement on the major scales of the SCL-90 and SDI, including the 17-item HRS, in a 49-patient sample receiving AMI 150 mg/day. A threshold level of AMI, NT or total TCA below which drug response was not apparent also were not detected.

The NT/AMI plasma ratio was computed for the 48 patients in whom 4-week values were available. As shown in figure 1, there was no correlation of NT/AMI ratio to improvement in the 17-item HRS scale from the SDI. The slope of the HRS improvement score vs. NT/AMI plot was 0.38 with a y intercept of 11.21 and a correlation coefficient (r) = 0.04 (NS).

TCA and 10-OH-TCA levels were analyzed in 18 of the most recent AMI-treated patients. Mean steady state levels after 2, 4 and 6 weeks of AMI treatment are shown in figure 2. It can be seen that mean 10-OH-NT concentrations are highest in plasma of patients treated with amitriptyline and 10-OH-AMI levels lowest. AMI and NT levels are slightly less than 10-OH-NT levels on the average and nearly equivalent to each other (NT/AMI ratio = 1.05). Neither 10-

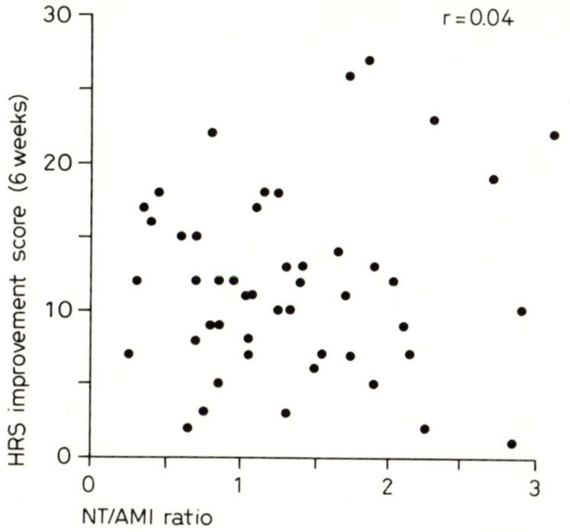

Fig. 1. Clinical improvement on 17-item Hamilton Rating Scale (HRS) for depression after 6 weeks of amitriptyline 150 mg/day is shown in relation to the demethylation ratio, nortriptyline (NT)/amitriptyline (AMI) in plasma.

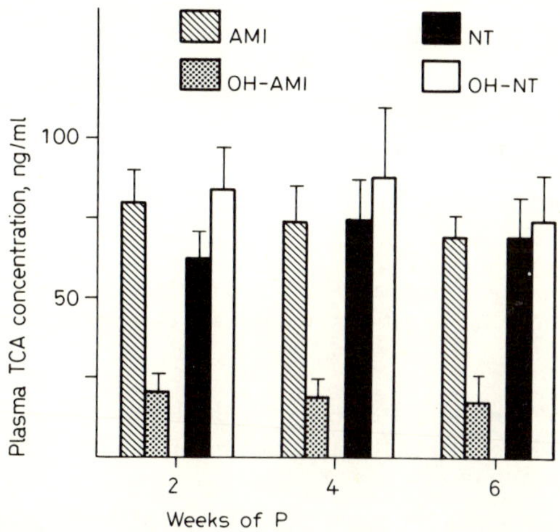

Fig. 2. Mean plasma concentration of amitriptyline (AMI), 10-hydroxy-amitriptyline (10-OH-AMI), nortriptyline (NT), and 10-hydroxy-nortriptyline (10-OH-NT) are shown after 2, 4 and 6 weeks of treatment with AMI 150 mg/day (n = 16).

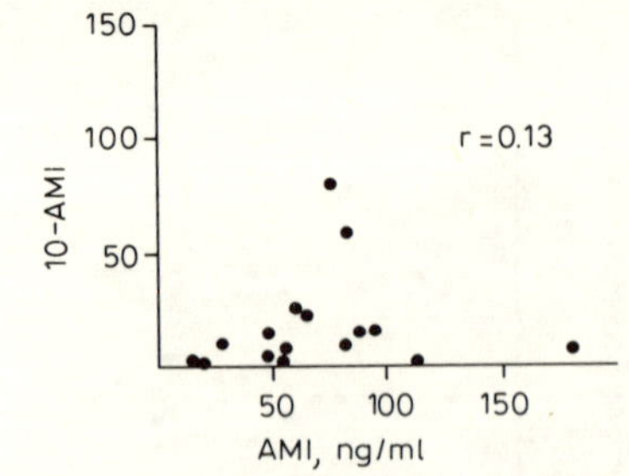

Fig. 3. The relation of plasma 10-hydroxy-amitriptyline (10-AMI) to amitriptyline (AMI) levels in ng/ml is shown for 16 patients after 4 weeks of treatment with AMI 150 mg/day.

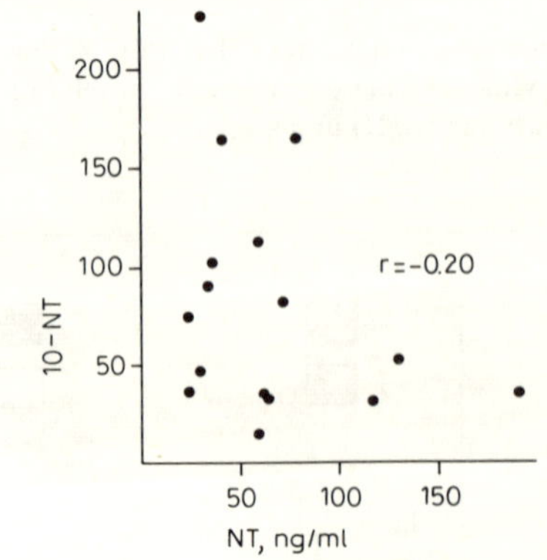

Fig. 4. The relation of plasma 10-hydroxy-nortriptyline (10-NT) to nortriptyline (NT) levels in ng/ml is shown for 16 patients after 4 weeks of treatment with AMI 150 mg/day.

hydroxylated metabolite correlated with the level of its respective parent compound (r = 0.134 for 10-OH-AMI vs. AMI; r = 0.204 for 10-OH-NT vs. NT) (fig. 3, 4).

Improvement scores on major SCL-90 and SDI scales were examined for associations with plasma levels of each compound as well as the sum of the 4 compounds (total TCA). As shown in table I. there were few significant correlations of amelioration scores with any of the 4 TCA compounds or total TCA. In this preliminary sample of 16 patients only 10-OH-NT levels were found to correlate

Table I. Relation of tricyclic antidepressant (TCA) plasma levels and drug response to amitriptyline[1]

% improvement	Kendall tau (n = 64)		Kendall's tau (n = 16)		
	AMI	NT	OH-AMI	OH-NT	total TCA
SCL-90					
Depression	–0.08	–0.02	0.00	–0.22	–0.09
Anxiety	0.02	–0.06	0.24	–0.01	
Total score	–0.03	–0.06	0.13	–0.18	–0.20
SDI					
Depression	0.00	0.07	–0.11	–0.35*	–0.03
Anxiety	–0.05	–0.11	0.00	–0.15	
17-item Ham	0.07	0.04	–0.16	–0.35*	0.01

* $p < 0.05$.

[1] Plasma concentrations (ng/ml) of amitriptyline (AMI), nortriptyline (NT), 10-hydroxy-amitriptyline (OH-AMI), 10-hydroxy-nortriptyline (OH-NT), and sum of all 4 compounds (total TCA) after 4 weeks of amitriptyline treatment (150 mg/day) are correlated (Kendall's tau) with relative improvement at 6 weeks on Symptom Check List (SCL-90) and Structured Depressive Interview (SDI) symptoms scales.

Table II. Correlations of plasma TCA and metabolite levels with age[1]

	Kendall's tau		
	2 weeks (n = 18)	4 weeks (n = 15)	6 weeks (n =15)
AMI	0.40**	0.15	0.33*
10-OH-AMI	0.39**	0.28	0.15
NT	0.25	–0.05	0.06
10-OH-NT	0.39**	0.31	0.40*

* $p < 0.05$; ** $p < 0.01$.

[1] Plasma concentrations (ng/ml) of amitriptyline (AMI), nortriptyline (NT), 10-hydroxy-amitriptyline (10-OH-AMI), and 10-hydroxy-nortriptyline (10-OH-NT) at steady state after 4 weeks of treatment with AMI 150 mg/day are correlated (Kendall's tau) with patient age.

negatively with improvement in two Depression scales of the SDI. Plasma TCA concentrations did not relate to side effects for any of the compounds, nor to global improvement ratings.

Because significant age correlations with steady state TCA levels have been previously reported in amitriptyline-treated patients [25], we examined the relationship of age to TCA and metabolite levels in these patients. We found a consistent positive age correlation (Kendall's tau) for AMI and 10-OH-NT (table II). However, steady state 10-OH-AMI and NT levels did not correlate with age.

Discussion

It has been a major disappointment that plasma antidepressant assays have not proven to be of more convincing clinical value. Many investigators would confine plasma TCA monitoring to selected cases of suspected noncompliance or treatment-resistant patients [26, 27]. The weak correlations demonstrated in the positive studies are not impressive. For such a correlation to have meaningful predictive value, the correlation coefficient (or other measure of association) should exceed a minimum of 0.5 and preferably 0.75, or there should be a clear-cut threshold level which defines a lower limit of therapeutic range. Those studies reporting positive findings for TCAs have not found sufficiently high measures of association. The more recent negative TCA studies have failed to detect either correlation of drug level with improvement or a lower threshold of plasma concentration below which therapeutic response does not occur in patients receiving standard TCA dosages.

In reanalysis of our AMI-NT data to compute a demethylation (NT/AMI) ratio, we find no relation of this ratio to clinical improvement on any rating scales including the HRS (fig. 1). These negative results are consistent with *Coppen and RamaRao*'s [28] reanalysis of their data. In the study of *Coppen* et al. [16], their 18 recovered patients had a mean NT/AMI ratio of 0.85 ± 0.1, compared to 16 nonresponders with a ratio of 1.11 ± 0.15. Thus, neither study confirmed the initial report of *Jungkunz and Kuss* [18] who suggested that demethylation ratio relates to clinical outcome.

The demonstration of a pharmacological profile of activity of 10-OH-AMI and 10-OH-NT similar to that of the parent TCA compounds coupled with the development of practical and sensitive assays for the substances [29], fostered the hope that assay of all active compounds present in significant concentrations in plasma and other tissues would at least reveal some meaningful relationship. Although *Ziegler* et al. [30] found 10-OH-NT to be highly correlated with plasma NT ($r = 0.75$, $n = 36$, $p < 0.001$), others have failed to replicate this. *Garland* et al. [29] reported $r = 0.25$ for AMI and 10-OH-AMI, and $r = 0.01$ for NT and 10-OH-NT. Similarly, we fail to find such a correlation (fig. 3, 4).

Since the hydroxylated metabolites are probably not correlated with the parent compounds and since 10-OH-NT may be the TCA present in plasma in the greatest concentrations, it was hoped that assaying 10-OH-NT or all 4 compounds might bring out meaningful and useful drug level-clinical improvement relationships. Unfortunately, this has not been the case in our study so far (table I). Further investigation in a wider sample of patients is required to conclusively establish this point.

Another possible explanation for the failure of plasma TCA levels to relate to clinical effect has been suggested in the work of *Potter* et al. [10]. In tissue distribution studies of imipramine at steady state in the rat, these investigators found that brain concentration of 2-OH-IMI amounted to 1/3 of IMI levels. Since they also found this hydroxy metabolite to be at least equipotent or possibly more potent than the parent compound, IMI in inhibiting norepinephrine reuptake in synaptosomes, it is possible that the hydroxy metabolite contributes significantly to the overall pharmacologic and therapeutic effects of imipramine. Again, further studies will be required in man to establish the relative CNS activities of these secondary and tertiary amines and their respective hydroxy metabolites.

Conclusions

Ongoing investigation of relationships between plasma concentrations of TCAs and therapeutic effect have tended to show at best a weak, or in many cases, no association between drug level and response. Most of the recent and larger sample size studies in amitriptyline-treated depressed patients have yielded negative findings in this regard.

In further analysis of our recent 49 outpatient clinical trial of amitriptyline, the demethylation (NT/AMI) ratio shows no correlation with improvement scores on several major rating scales. This is in agreement with *Coppen and RamaRao* [28] who also reanalyzed data from their 55-patient cooperative study which used amitriptyline 150 mg/day in depressed inpatients.

The 10-hydroxy metabolites of AMI and NT were assayed in plasma in a recent subsample of patients from our ongoing amitriptyline controlled clinical trial. Mean steady state plasma levels of 10-OH-NT exceeded those of both NT and AMI. Steady state levels of 10-OH-AMI were lowest of the compounds. The 10-hydroxylated metabolites of AMI and NT did not correlate with levels of its respective parent compound.

Levels of the 10-hydroxylated metabolites and total TCA levels (all 4 active tricyclic compounds) in plasma showed no relationships to clinical therapeutic effects or side effects in the amitriptyline-treated patients except for a negative correlation of 10-OH-NT with depression scales. Currently, there is little evidence of the clinical utility of antidepressant plasma levels in the routine managment of depressed patients in regulating dosage in individuals. Perhaps the major indications for TCA blood levels are to confirm significant noncompliance, especially in the patient who has proven refractory to standard treatments and doses.

References

1 Hammer, W.M.; Brodie, B.B.: Application of isotope derivative technique to assay of secondary amines: estimation of desipramine by acetylation of ^{3}H-acetic anhydride. J. Pharmac. exp. Ther. *157:* 503–508 (1967).

2 Biggs, J.T.; Holland, W.H.; Chang, S.; Hipps, P.P.; Sherman, W.R.: Electron beam ionization mass fragmentographic analysis of tricyclic antidepressants in human plasma. J. pharm. Sci. *65:* 261–268 (1976).

3 Cooper, T.B.; Allen, D.; Simpson, G.M.: A sensitive method for the determination of amitriptyline and nortriptyline in human plasma. Commun. Psychopharm. *2:* 105–116 (1976).

4 Amidsen, A.: Blood concentration of tricyclic antidepressants as a daily routine? A critical review from the doorstep of the clinical laboratory. Acta psychiat. scand. *61:* suppl. 280, pp. 262–280 (1980).

5 Kane, J.; Rifkin, A.; Quitkin, F.; Klein, D.F.: Antidepressant drug blood levels, pharmacokinetics and clinical outcome; in Progress in psychiatric drug treatment, pp. 136–157 (Bruner/Mazel, New York 1976).

6 Amsterdam, J.; Brunswick, D.; Mendels, J.: The clinical application of tricyclic antidepressant pharmacokinetics and plasma levels. Am. J. Psychiat. *137:* 653–662 (1980).

7 Burrows, G.D.; Davies, B.; Norman, T.; Maguire, K.; Scoggins, B.A.: Should plasma levels monitoring of tricyclic antidepressants be introduced in clinical practice? Commun. Psychopharma. *2:* 393–408 (1978).

8 Javaid, J.I.; Perel, J.M.; Davis, J.M.: Inhibition of biogenic amines uptake by imipramine, desipramine, 2 OH-imipramine and 2 OH-desipramine in rat brain. Life Sci. *24:* 21–28 (1979).

9 Bertilsson, L.; Mellstrom, B.; Sjoqvist, F.: Pronounced inhibition of noradrenal in uptake by 10-hydroxy-metabolites of nortriptyline. Life Sci. *25:* 1285–1292 (1979).

10 Potter, W.Z.; Calil, H.M.; Manian, A.A.; Zavadil, A.P.; Goodwin, F.K.: Hydroxylated metabolites of tricyclic antidepressants. Preclinical assessment of activity. Biol. Psychiat. *14:* 601–613 (1979).

11 Gruvstad, J.: Plasma levels of antidepressants and clinical response. Lancet *i:* 95–96 (1973).

12 Braithwaite, R.A.; Goulding, R.; Theano, G.; Bailey. J.; Coppen, A.: Plasma concentration of amitriptyline and clinical response. Lancet *i:* 1297–1300 (1972).

13 Ziegler, D.E.; CO, B.T.; Taylor, J.R.; Clayton, P.J.; Biggs, J.T.: Amitriptyline plasma levels and therapeutic response. Clin. Pharmacol. Ther. *6:* 795–801 (1976).

14 Kupfer, D.J.; Hanin, I.; Spiker, D.; Neil, J.; Coble, P.; Grau, T.; Nevar, C.: Amitriptyline plasma levels and clinical response in primary depression. Part II. Commun. Psychopharm. *2:* 441–450 (1978).

15 Montgomery, S.A.; McAully, R.; Rani, S.J.; Montgomery, D.B.; Braithwaite, R.; Dawlings, S.: Amitriptyline plasma level concentration and clinical response. Br. med. J. *i:* 123–124 (1979).

16 Coppen, A.; Ghose, K.; Montgomery, S.; Rama Rao, V.A.; Bailey, J.; Christiansen, J.;

Mikkleson, P.L.; Praag, H.M. van; Poel, F. van de, Minsker, E.J.; Kozulja, V.J.; Matussek, N.; Kungkunz, G.; Jorgensen, A.: Amitriptyline plasma-concentration and clinical effect. Lancet *i:* 63–66 (1978).

17 Liisberg, P.; Mose, H.; Amdisen, A.; Jorgensen, A.; Hopfner Peterson, H.E.: A clinical trial comparing sustained release amitriptyline (Saroten Retard) and conventional amitriptyline tablets (Saroten) in endogenously depressed patients with simultaneous determination of serum levels of amitriptyline and nortriptyline. Acta psychiat. scand. *57:* 426–435 (1978).

18 Jungkunz, G.; Kuss, H.J.: Amitriptyline and its demethylization rate. Lancet *ii:* 1263–1264 (1978).

19 Robinson, D.S.; Cooper, T.B.; Ravaris, C.L.; Ives, J.O.; Nies, A.; Bartlett, D.; Lamborn, K.R.: Plasma tricyclic drug levels in amitriptyline-treated depressed patients. Psychopharmacology *63:* 223–231 (1979).

20 Mendlewicz, J.; Linkowski, P.; Rees, J.A.: A double-blind comparison of dothiepin and amitriptyline in patients with primary affective disorders. Serum levels and clinical response. Br. J. Psychiat. *136:* 154–160 (1980).

21 Borga, O.; Garle, J.: A gas chromatographic method for the quantitative determiniation of nortriptyline and some of its metabolites in human plasma and urine. J. Chromat. *68:* 77–88 (1972).

22 Hamilton, M.: A rating scale for depression. J. Neurol. Neurosurg. Psychiat. *23:* 56–62 (1960).

23 Nies, A.; Robinson, D.S.; Lamborn, K.R.: The efficacy of the MAO inhibitor, phenelzine: Dose effects and prediction of response; in Neuropsychopharmacology, pp. 765–770 (Excerpta Medica, Amsterdam 1974).

24 Derogatis, L.R.: SCL-90 manual, 1rst ed. (Johns Hopkins University, Baltimore, 1977).

25 Nies, A.; Robinson, D.S.; Freidman, M.J.; Green, R.; Cooper, T.B.; Ravaris, C.L.; Ives, J.O.: Relationship between age and tricyclic antidepressant plasma levels. Am. J. Psychiat. *134:* 790–793 (1977).

26 Shader, R.I.; Greenblatt, D.J.: Clinical indications for plasma level monitoring of psychotropic drugs. Am. J. Psychiat. *136:* 1590–1591 (1979).

27 Hollister, L.E.: Monitoring tricyclic antidepressant plasma concentrations. J. Am. med. Ass. *241:* 2530–2533 (1979).

28 Coppen, A.; Rama Rao, V.A.: Amitriptyline and its demethylation-rate. Lancet *i:* 49 (1979).

29 Garland, W.A.; Muccino, R.R.; Min, B.H.; Cupani, J.; Fann, W.E.: A method for the determination of amitriptyline and its metabolites nortriptyline, 10-hydroxy amitriptyline, and 10-hydroxy nortriptyline in human plasma using stable isotope dilution and gas chromatography-chemical ionization mass spectometry (GC-CIMS). Clin. Pharmacol. Ther. *25:* 844–856 (1980).

30 Ziegler, V.A.; Fuller, T.A.; Biggs, J.T.: Nortriptyline and 10-hydroxy nortriptyline plasma concentrations. J. Pharm. Pharmac. *28:* 849–850 (1976).

D.S. Robinson, MD, Department of Pharmacology Doctors' Memorial Building, Marshall University School of Medicine, Huntington, WV 25701 (USA)

Adv. biol. Psychiat., vol. 7, pp. 208–217 (Karger, Basel 1981)

Lithium and Antidepressants in the Prophylaxis of Depression[1]

Iain Glen

MRC Brain Metabolism Unit, Thomas Clouston Clinic, Edinburgh, Scotland

The benefits of continued drug treatment for depression after the disappearance of the symptoms of an episode remains a subject of controversy. I would like (1) to examine the results of recent trials of such continued treatment in which lithium has been compared with antidepressant medication; (2) to present the first report of the MRC multi-centre trial, a double-blind study comparing the effects of lithium, amitriptyline and a placebo in patients exposed to the medication for a period of 3 years, and finally (3) to discuss the implications of these results for current psychiatric practice.

The efficacy of tricyclic antidepressants in the treatment of depressive illness is established. Evidence for this was given by the Medical Research Council trial of 1965 and was the subject of a review in the same year by *Wechsler* et al. [1965]. The Medical Research Council trial of 1973 [*Mindham* et al., 1973] established the value of continued treatment with imipramine or amitriptyline in recurrent unipolar depressive illness for 6 months following recovery from an episode. In this study of 92 patients who entered the trial, only 22% on active treatment (amitriptyline or imipramine) relapsed compared with 50% of patients receiving placebo during the 6-month trial period. With the introduction and widespread use of lithium in the early 1970s, many patients with recurrent unipolar depressive illness began to receive lithium although results of trials of the value of lithium in unipolar depressive illness were only limited, i.e. few patients were studied. Some psychiatrists, however, felt that where patients had clear-cut

[1] The multi-centre drug trial of lithium and amitriptyline in unipolar depressive illness described in this report was carried out by the Medical Research Council under the auspices of the Drug Trials Sub-Committee. A full list of the names of contributors is given in the text. A major contribution in the design of the trial came from Prof. *Michael Shepherd* and initially also from Prof. *G.W. Ashcroft.* Throughout the trial considerable advice was obtained from Dr. *Ian Sutherland* and from Dr. *Tony Johnson* of the MRC Statistics Unit and the latter has been responsible for the statistical analysis.

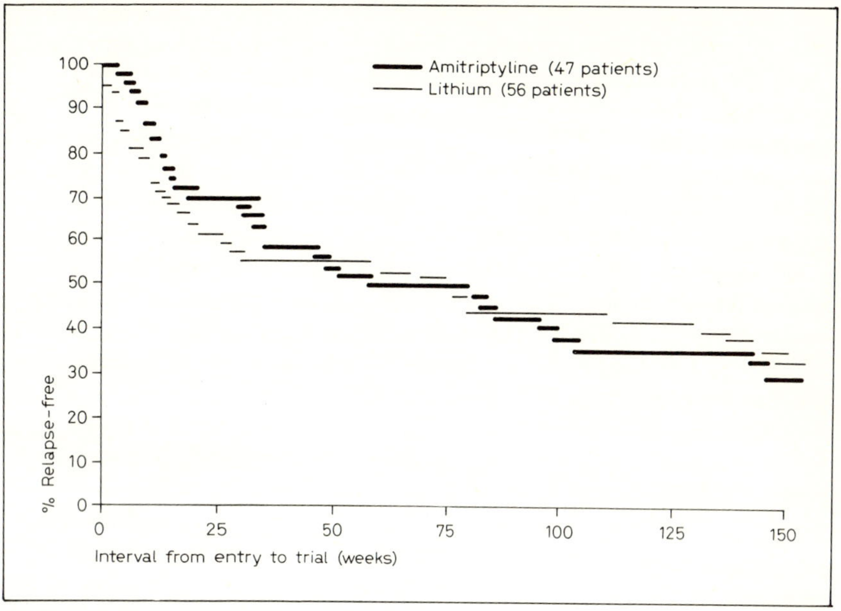

Fig. 1. Actuarial percentages of relapse free patients in Group I (patients receiving lithium or amitriptyline).

Table I. Studies comparing lithium and placebo in unipolar depression

Reference	Number of subjects	Comment	
		lithium	placebo
Melia [1970]	4	too few for analysis	
Coppen et al. [1971]	26	morbidity index 0.08	morbidity index 0.54
Cundall et al. [1972]	5	too few for analysis	
Hullin et al. [1972]	23	78% relapse rate	
Fieve et al. [1975]	28	0.59 episodes/patient year	1.60 episodes/patient year

episodes, often psychotic, it was reasonable to extrapolate from results of studies of bipolar illness which showed a clear preference for lithium. Growing anxiety about the potential toxicity of lithium has, however, led many psychiatrists to review their therapeutic practice with regard to the treatment of unipolar patients with lithium and to await the results of new studies. Such anxiety about the potential effects of lithium on the kidney may be seen in the change in prescribing practice in Edinburgh which took place following the publication of the study of *Hestbech* et al. [1977] (fig. 1). This information was obtained from a registered population of patients in which the total number of patients prescribed lithium in the City of Edinburgh and its environs are constantly under review. The register indicates an abrupt change in prescribing practice for lithium in 1978 in Edinburgh with a recovery to previous prescribing habits following the reassuring reports of 1979 [*Lancet,* 1979].

A number of controlled studies of the continued treatment of unipolar depressives following recovery have previously been reported in addition to those of the Medical Research Council. In the United States *Prien* et al. [1973] compared the effects of lithium, imipramine and placebo in 78 patients following in the trial for a period of a year. The results showed both lithium and imipramine to be superior to placebo, the relapse rate in the lithium group being 41%, in the imipramine group 29% and in the placebo group 85%. A number of other trials had included unipolar patients as well as bipolar patients and in these studies significant differences between lithium and placebo in the unipolar groups were found. The numbers of patients in some of these trials, however, were small and in others methodological problems of trials design were not overcome. The results of these studies are summarised in table I. None can be said to provide a conclusive result. In 1975 the new dimension of control of plasma level of antidepressant medication as well as lithium began to be reported.

Coppen et al. [1976] also compared maprotiline and lithium and found a significantly lower morbidity index on lithium as compared with maprotiline. In a further study *Coppen* et al. [1978] compared lithium with mianserin and found no significant difference between the two treatments in regard to the morbidity index. However, there were significantly more admissions to hospital for treatment in the mianserin group.

The most recent study, a preliminary report from the United States by *Quitkin* et al. [1978], used a double-blind placebo technique and found lithium superior to both imipramine and placebo in the unipolar patients studied. The numbers in this study at the present time are small. The results of these studies are summarised in table II.

The Medical Research Council conducted an extensive review of the current status of lithium as regards its toxicity and therapeutic efficacy in 1975. A year later in the United States a similar exercise was carried out and both bodies came to the same conclusion, namely that for unipolar depressive illness the evidence

Table II. Studies comparing lithium and antidepressants in the prophylaxis of unipolar depressive illness

Reference	Number of subjects	Comment	
		lithium	antidepressant
Coppen et al. [1976] Lithium vs. maprotiline	30 at start 21 at finish	morbidity index 0.11	morbidity index 0.24
Coppen et al. [1978] Lithium vs. mianserin	26	morbidity index 0.229 significantly more admission for mianserin group	morbidity index 0.231
Quitkin et al. [1978] Lithium vs. imipramine	21 (approx 5 per group)	40% relapse	80% relapse

Table III. Participants in the trial

Cardiff

Regional organiser: Dr. *D.M. Shaw*, Biochemist Mr. *Tidmarsh*, Pharmacist Mr. *Mitchel*, Psychiatrists Drs. *Michaelakeas*, *Karajgi* and *Gray*

Chichester

Regional organiser: Dr. *P. Sainsbury*, Biochemist Dr. *Lord*, Pharmacist Mr. *Frostick*, Psychiatrist Dr. *Resek*, Psychologist Mr. *Kennard*

Dundee

Regional organiser: Dr. *G.J. Naylor*, Biochemist Dr. *Moodie*, Pharmacist Miss *Miller*, Psychiatrist Dr. *Hopwood*

Edinburgh

Regional organisers: Dr. *J.B. Loudon* and Dr. *J.E. Christie*, Biochemist Dr. *Reading*, Pharmacist, Mrs. *Gillies*, Psychiatrists Drs. *Strachan*, *Lipiett*, *Fisk*, *Sashidharan*, *Williamson*, *Blackwood*, *Waring*, *Miller*, *Backett*, *Dutta*, *George*, *Ashcroft*, *Eccleston Jacques*, *Barnes*, *Cox*, *Horgan*, *Abousaleh*, *Naya*, *Van Lare*, *Affleck*, *Rix*, *Kolman*, *Adamson*, *Waite*, *Thomas*, *Fairburn*, *Tyre*, *Obeiwi* and *Parry*

Frimley Park

Regional organiser: Dr. *D. Hassall*, Biochemist Dr. *Alston*, Pharmacist Mrs. *Hutchinson*

Glasgow

Regional organiser: Dr. *E.P. Worrall*, Biochemist, Dr. *Thompson*, Pharmacist Mrs. *Waterman*

Leeds

Regional organiser and Biochemist: Dr. R.P. *Hullin*, Pharmacist Mr. *Frayne*, Psychiatrists Drs. *Jerram*, *Coley*, *Srinivasan*, *Wallis*, *Jardine*, *Snaith*, *Alker* and Dr. *Oliver* of Thameside General Hospital

Sheffield

Regional organiser: Prof. *F.A. Jenner*, Biochemist Dr. *Baillie*, Pharmacist Mr. *Ryan*, Psychiatrist Dr. *Paschalis*

Southampton

Regional Organisers: Dr. *P.J. Tyrer* and Dr. *G. Edwards*, Biochemist Dr. *Middleton*, Pharmacist Mrs. *Tyler*, Psychiatrists Drs. *Trotter*, *Gill*, *Baldwin*, *Steinberg*, *Stewart*, *Keitch* and *Rastagi*

was still insufficient to conclude that lithium was of benefit. In the United Kingdom a multi-centre trial of lithium comparing its effects with amitriptyline and placebo was therefore undertaken. Preliminary results from this trial are now emerging and it is as the co-ordinator of the trial that I wish to make some observations. I would like to emphasise that these results are provisional and the overall conclusions have not been reviewed as yet by the Medical Research Council. I have, however, been given the authority to make a personal statement at the present time.

I would like first to emphasise the considerable effort involved in a trial of this kind and to list all those who made a major contribution (table III).

Table IV. Entry criteria

Overall criteria
The subjects should be between 25 and 65 years of either sex
The subjects should be patients entering a hospital with a primary depressive illness or should be attending a psychiatric out-patient department
There should be no previous history of mania
Subjects should have no physical disability sufficient to preclude treatment with either lithium or amitriptyline
Subjects should have recovered from the index episode prior to entry to the trial

Definition of a primary depressive illness
The episode should have lasted at least 1 month and should have had persistent and continuing alteration in mood during that time [Medical Research Council, 1965]
The patients should fulfil standard criteria for a major depressive illness [*Spitzer* et al., 1978]

Table V. Trial design and follow-up

Index episode	PSE Hamilton Rating Scale MRC rating form previous history of episodes
Entry to trial	random assignment to active drug or placebo by quota allocation in age and sex groups
8-weekly assessment or on relapse or withdrawal	psychiatric ratings toxic side effects plasma level of drug
6-monthly physical assessment	haematology plasma electrolytes and thyroid status ECG creatinine clearance
3 years	

Table VI. Equivalent units of plasma lithium and amitriptyline

Reported result in units of trial drug	Equivalent plasma lithium, m*M*/1 $\frac{\text{units}}{100}$	Equivalent plama amitriptyline + nortryptline, ng/ml units $\times \frac{3}{2}$
50	0.50	75
80	0.80	120
100	1.00	150
120	1.20	180
140	1.40	210

Method

We aimed at meeting a number of methodological objections which had emerged in previous studies in the trial design. The design strategy had, however, to be compatible with the needs of current clinical practice in the treatment of depression in 1975. Patients were divided into two groups. Those patients with more than one episode of depression prior to the index depression leading to entry to the trial received lithium or amitriptyline. Patients with only one episode of depression prior to the index depression received lithium, amitriptyline or placebo. The entry criteria to the trial are shown in table IV. The overall trial design and the follow-up assessment are shown in table V. All patients received tablets similar in outward appearance. Blood levels of the trial drug were monitored at 8-weekly intervals or as often as the clinician demanded in each patient. Drug levels were reported to the clinician as units of trial medication, each unit being equivalent to a drug standard: i.e., for a patient on lithium 100 units were equivalent to 1 mmol/l of plasma lithium; for those on amitriptyline 100 units were equivalent to 150 ng/ml of amitriptyline plus nortriptyline. Toxic levels were reported urgently to the clinician at a cut-off point of 120 units, i.e. 1.2 mmol/l of plasma lithium or 180 ng/ml combined amitriptyline plus nortriptyline. The equivalent units of comparison are shown in table VI.

Results

The preliminary analysis of results has been confined to the effect of the decision by the clinician to allocate the patient to one or other treatment. The analysis has been carried out by Dr. *Johnson* of the MRC Statistics Unit using the log-rank test which assesses the relapse-free survival rate in each group of patients. These preliminary analyses are shown diagrammatically in figure 2.

The overall results at this preliminary stage of analysis suggest that both lithium and amitriptyline are superior to a placebo but that differences between the two active treatments are minimal.

Discussion

The results of the Lithium/Amitriptyline Trial are in keeping with the studies of *Prien* et al. [1973] who found that there was no significant difference between the therapeutic effects of lithium and imipramine in the treatment of unipolar depressive illness but that some patients on imipramine and none on lithium developed *manic* symptoms (table VII). 2 patients in the MRC trial developed a *manic* episode on the amitriptyline regime but no patients on lithium showed manic symptoms. The relapse-free survival at 6 months is similar to the previous results of continuation treatment [*Mindham* et al., 1973] (table VIII). The implications of the *Prien* et al. [1973] study and the MRC Lithium/Amitriptyline Trial taken together are that two preparations with apparently very different pharmacological properties produce similar treatment results. For the

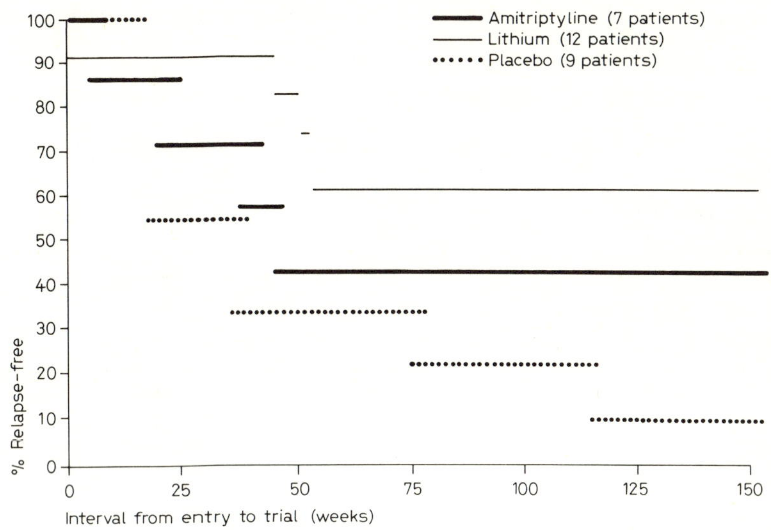

Fig. 2. Actuarial percentages of relapse free patients in Group II (patients receiving lithium, amitriptyline or a placebo).

future, the question of the relationship between therapeutic effect and toxicity must surely be further examined. The results of the toxicity assessment are clearly of critical importance in considering the appropriate drug regime for an individual patient and the analysis of toxicity together with an examination of relative morbidity, i.e. continuing symptomatology in the two groups of patients will be reported later. 3 patients died during the study, 1 patient on amitriptyline had a suicidal death, 1 patient on amitriptyline a cardiac death, and 1 patient on lithium a cardiac death.

Coppen et al. [1976, 1978] concluded from results of double-blind comparisons between lithium and maprotiline and lithium and mianserin that lithium was superior to both treatments and that lithium is the choice for continuation treatment in unipolar depressive illness. The results of the present study do not support this conclusion. Some causes of the differences between this study and the comparisons between lithium and maprotiline and mianserin can be suggested: (1) patients in these studies were already receiving lithium and attending a lithium clinic and the sample may have been biased in favour of lithium responders; (2) the reference drug amitriptyline used in the present study may be superior to mianserin and maprotiline; (3) the reference drug in the present

Table VII. Percentage of patients remaining free from relapse at 2 and 3 years

	Prien et a. [1973] (n = 78)	Medical Research Council [1980] (n = 131)			
		2 years	3 years		
Lithium	37	45	64	32	64
Amitriptyline/imipramine	44	35	43	30	43
Placebo	8	–	22	–	22

Table VIII. Percentage of patients remaining free from relapse at 6 months

	Mindham et al. [1973] (n = 92)	Medical Research Council [1980] (n = 131)	
		I	II
Lithium	–	60	91
Amitriptyline/imipramine	80	70	72
Placebo	52	–	45

study was adjusted to give an optimal plasma level; (4) assessment of patients by a morbidity index in the report of *Coppen* et al. [1976] rather than by relapse rate as in the present study favoured lithium, i.e. if as in this study patients re-entered the trial on relapse, the results might favour lithium, since some clinicians believe that lithium may take at least a year to stabilise the illness.

Some of these aspects of continuation treatment will be examined in a further report. In this preliminary communication it should be emphasised that the question asked was of the effect of a therapeutic decision to place a patient either on lithium, amitriptyline or placebo. The answer given by this study to this question in terms of relapse is that both treatments are superior to placebo but that lithium is not significantly better than amitriptyline.

References

1 Coppen, A.; Ghose, K.; Rao, R.; Bailey, J.; Peet, M.: Mianserin and lithium in the prophylaxis of depression Br. J. Psychiat. *133:* 206–210 (1978).

2 Coppen, A.; Montgomery, S.A.; Gupta, R.K.; Bailey, J.E.: A double-blind comparison

of lithium carbonate and maprotiline in the prophylaxis of the affective disorders Br. J. Psychiat. *128:* 479–485 (1976).

3 Coppen, A.; Noguera, R.; Bailey, J.; Burns, B.J.; Swani, M.S.; Hare, E.H.; Gardner, R.; Maggs, R.: Prophylactic lithium in affective disorders. Lancet *ii:* 275 (1971).

4 Cundall, R.I.; Brooks, P.W.; Murray, L.G.: Controlled evaluation of lithium prophylaxis in affective disorders. Psychol. Med. *3:* 308–311 (1972).

5 Fieve, R.R.; Dunner, D.L.; Kumberachi, T.: Lithium carbonate in affective disorders. IV. A double-blind study of prophylaxis in unipolar recurrent depression. Archs. gen. Psychiat. *32:* 1541–1544 (1975).

6 Hestbech, J.; Hansen, H.E.; Amdisen, A.; Olsen, S.: Chronic renal lesions following long-term treatment with lithium Kidney int. *12:* 205–213 (1977).

7 Hullin, R.P.; McDonald, R.; Allsopp, M.N.: Prophylactic lithium in recurrent affective disorders Lancet *i:* 1044–1046 (1972).

8 Lancet: Editorial Lancet *ii:* 1056–1057 (1979).

9 Medical Research Council: Clinical trial of the treatment of depressive illness. Br. med. J. *i:* 881–886 (1965).

10 Melia, P.I.: Prophylactic lithium. A double-blind trial in recurrent affective disorders. Br. J. Psychiat. *116:* 621–624 (1970).

11 Mindham, R.H.S.; Howland, C.; Shepherd, M.: An evaluation of continuation therapy with tricyclic antidepressants in depressive illness. Psychol. Med. *3:* 5–17 (1973).

12 Prien, R.F.; Klett, J.; Caffey, E.M., Jr.: Lithium carbonate and imipramine in prevention of affective episodes. Archs. gen. Psychiat. *29:* 420–425 (1973).

13 Quitkin, F.; Rifkin, A.; Kane, J.; Ramos-Lorenzi, J.R.; Klein, D.F.: Prophylactic effect of lithium and imipramine in unipolar and bipolar II patients. A preliminary report. Am. J. Psychiat. *135:* 570–572 (1978).

14 Spitzer, R.L.; Endicott, J.; Robins, E.: Research diagnostic criteria (RDC) for a selected group of functional disorders (Biometrics Research, N.Y. State Psychiatric Institute, New York 1978).

15 Wechsler, H.; Grosser, G.H.; Greenblatt, M.: Research evaluating antidepressant medications on hospitalized mental patients. A survey of published reports during a five-year period. J. nerv. ment. Dis. *141:* 231–239 (1965).

I. Glen, MD, Southern District – Highland Health Board, Craig Dunain Hospital, Iverness, IV3 6JU (Scotland)

Adv. biol. Psychiat., vol. 7, pp. 218–229 (Karger, Basel 1981)

Clinical Indications for a Prophylactic Treatment of Depression

Jules Angst

Psychiatric University Clinic, Research Department
(Head: Prof. Dr. med. *Jules Angst*), Zürich, Switzerland

During the last 20 years we have followed-up a considerable number of patients with affective disorders in our Psychiatric University Clinic in Zürich. These patients hardly ever received lithium and, therefore, the observed episodes should reflect the natural history of the disorder.

The patient sample has already been described in detail [*Angst,* 1980]. Some of its characteristics are mentioned in table I. The sample consists of 159 unipolar depressive, 95 bipolar manic-depressive, and 150 schizoaffective patients. At the last follow-up in 1975, the average age was approximately 60 years. The patients have been observed during 19–27 years since the onset of illness.

This study is based on these observations. Our data show that not every patient suffering from an affective disorder requires lithium treatment. In this study our data will be used in order to optimize the criteria for selecting the right patients. Different solutions for an optimal lithium indication will be compared in a simulation study. As a second problem, the question will be dealt with as to how long lithium prophylaxis should be maintained in old age?

The decision whether to start lithium prophylaxis or not will be examined from 2 aspects: – considering the *diagnosis,* and – considering the *individual previous history* of the illness.

Importance of Diagnosis with Regard to Lithium Prophylaxis

This paper is restricted to affective disorders which respond to lithium prophylaxis. Within the subgroups of affective disorders we know that bipolar and schizoaffective psychoses show a similarly high relapse risk, whereas unipolar depression has a much more favourable course (fig. 1). The better prognosis in unipolar depression is to a certain extent biased by the relatively high suicide rate (10%) in this group.

Table I. Description of the sample

	Unipolar (n = 159)	Bipolar (n = 95)	Schizoaffective (n = 150)
Male	39 (24.5%)	37 (38.9%)	41 (27.3%)
Female	120	58	109
Age, years	64.3 ± 13.9	61.0 ± 13.2	58.9 ± 13.7
Years of observation	19.0 ± 11.4	26.4 ± 11.9	27.0 ± 12.8
Age at first manifestation	45.3 ± 16.6	34.7 ± 13.9	31.8 ± 12.9

Table II. Number of episodes

Number of Episodes	Unipolar			Bipolar			Schizoaffective		
	n	%		n	%		n	%	
1	24	15	49 (1–3)	–	–	8 (1–3)	5	3	13 (1–3)
2	35	22		2	2		4	3	
3	19	12		6	6		10	7	
⩾ 4	81	51		87	92		131	87	
Total	159			95			150		
Median	4.0			9.0			7.0		
1st + 3rd quartile	2.0–6.0			5.0–16.0			5.0–12.0		

It is obvious that long-term prophylactic medication is required only if a disorder is really recurrent. The question of the total number of episodes observed during our study is, therefore, relevant. Table II shows that single episodes practically do not occur among bipolar and schizoaffective psychoses. Among cases with unipolar depression, however, we observed 15% with a single episode, 22% with two episodes, and 12% with three episodes only, during an average period of 19 years. Assuming that prophylaxis is certainly not indicated if a patient suffers from three or less episodes during at least 20 years, 49% of the unipolar group, 8% of the bipolar, and 13% of the schizoaffective group would never be in need of lithium treatment.

Individual Previous History with Regard to Lithium Prophylaxis

In order to choose the right patients for long-term prophylaxis within these diagnostic groups, we have to consider the individuals' previous history; namely, the number of previous episodes, the length of cycles (from the beginning of one

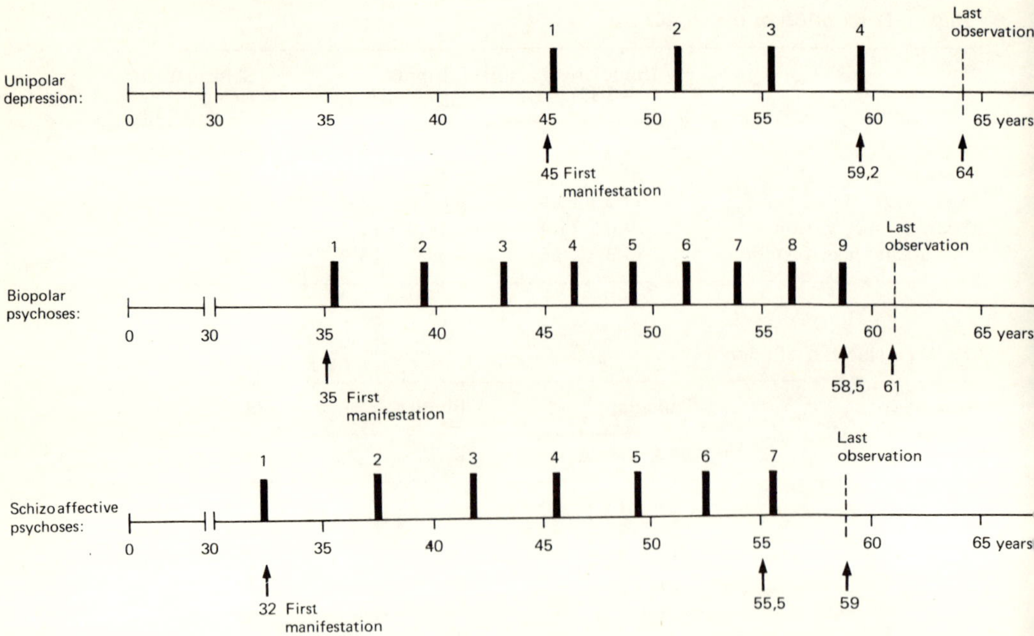

Fig. 1. Occurrence of episodes in unipolar depression, bipolar, and schizoaffective psychoses.

episode to the beginning of the next), the severity of the episodes, the personal and social consequences of the disorder, and the degree of remission. Up to now there is no strict rule on how to weight up these factors in order to make a clear decision whether lithium treatment should be started or not.

Relapse Risk over 5 Years

The first step is to determine if the risk of recurrence is high enough to justify long-term prophylactic medication. It is obvious that patients without a relapse risk should not be treated prophylactically. Among the results of our followed-up patients who did not receive prophylactic treatment we found a substantial number without relapse. 37% of the unipolar, 15% of the bipolar, and 18% of the schizoaffective group have not experienced any further episodes after the initial one, and, consequently, have not required lithium treatment.

According to our experience we think that the relapse risk over a 5-year follow-up period should be at least two episodes. Applying this criterion, 38% of

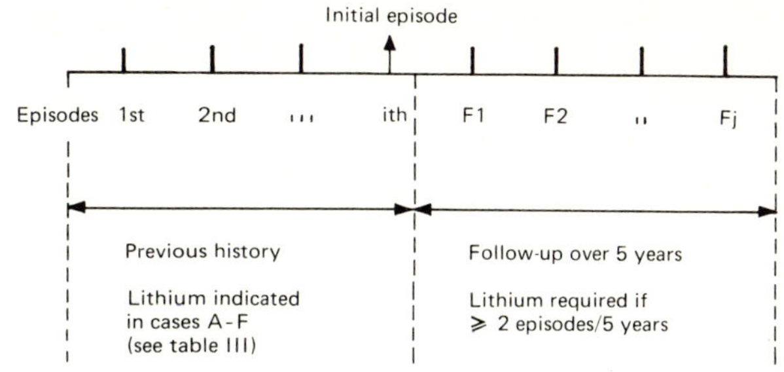

Fig. 2. Previous history and follow-up history.

the unipolar depressed, 66% of the bipolar and 60% of the schizoaffective patients would require lithium treatment. But our problem still is, how can we recognize these 38 or 66% of the patients who show a clearly recurrent affective disorder, and exclude those with good prognosis who require no long-term medication at all?

In order to look for an optimal solution we carried out a simulation study on our empirical data. The available information can be split into previous history including the initial episode and follow-up history (fig. 2). Lithium prophylaxis is considered justified if the patient had at least two further episodes during the 5-year follow-up period. From this, we can validate some of the criteria used in scientific studies and in practice for choosing patients for lithium treatment. These criteria rely mainly upon the number of previous episodes and are listed in table III. Criteria A, B and C are frequently mentioned in research projects for selecting more or less homogeneous groups with a very high risk of recurrence. On the other hand, we have also tested criteria D, E, and F which are less strict as they include a much larger proportion of patients to be treated with lithium.

The basis of the statistical analysis is shown in table IV which reproduces the decision pattern for a lithium indication based on the previous history and its validity in view of the follow-up history. The optimal solution lies on the diagonal (a–d), maximizing the 'true positive' decisions (a), and the 'true negative' decisions (d). Too many 'false positive' decisions (b) mean overtreatment of a patient sample, too many 'false negative' decisions (c) mean undertreatment of the patient sample.

The details of the analysis are given in table V considering 'specificity', 'sensitivity', and the so-called 'false negativity' and 'false positivity', and, furthermore, the 'overall efficiency'. The latter reflects the percentage of 'true positives'

Table III. Criteria for lithium treatment based on previous history episodes (the initial episode is included in the number of episodes)

A	yearly ⩾ 1 episode	over 3 years
B	yearly ⩾ 1 episode	over 2 years
C	yearly ⩾ 2 episodes	over 1 year
D	⩾ 2 episodes	over 3 years
E	⩾ 2 episodes	over 4 years
F	⩾ 2 episodes	over 5 years

Table IV. Decision pattern for an indication for lithium

Previous history criteria	Follow-up criteria	
	fulfilled	not fulfilled
Fulfilled	'true positive' (a)	'false positive' (b)
Not fulfilled	'false negative' (c)	'true negative' (d)

and 'true negatives' together. The optimal solutions are shown in a square: i.e. F for unipolar depression, E for bipolar depression, and D for schizoaffective psychoses.

Figure 3 shows the frequency of overtreatment ('false positive' decisions in proportion to all negatives). We call this percentage 'false positivity'. It represents all patients who do not need lithium and yet received it. We can see that research criteria A, B and C show very good results as there is hardly any overtreatment. On the other hand applying the less strict criteria and thereby including more patients for treatment, we arrive at a rate of about 20% overtreatment.

The frequency of undertreatment ('false negativity') is given in figure 4 which shows the opposite picture. Criteria A, B and C include an undertreatment of about 80% which means that the majority of the patients who require lithium do not receive it. The less stringent criteria D, E and F are better in this respect, but they still exclude about 50% of the patients who are in need of lithium.

In practice, it is not only the 'overall efficiancy', i.e. the sum of 'true positives' and 'true negatives' alone that are relevant for the optimal solution, but also the ratio of overtreatment to undertreatment. From a practical point of view we assume that slight overtreatment is preferable to major undertreatment of a patient group. Considering these two aspects we come to a compromise with a practical recommendation for selecting patients for lithium prophylaxis which is illustrated in table VI.

Table V. Statistical analysis of the validity of different criteria for selecting patients

Diagnosis and Criteria for selection		TP % (a)	FP % (b)	FN % (c)	TN % (d)	1	2	3	4	5	Kappa coeff.	Z	p
Unipolar	A	5	0	33	62	67	100	65	88	0	14.7	1.339	n.s.
depression	B	6	2	32	60	66	80	65	84	3	16.3	1.505	n.s.
(n = 128)	C	13	1	25	61	74	94	71	65	1	38.0	3.626	0.001
	D	16	8	22	54	70	68	71	57	13	32.5	3.252	0.001
	E	17	10	21	52	69	65	71	55	15	31.5	3.19	0.01
	F	19	12	19	50	69	62	73	*49*	*19*	*33.2*	3.438	0.001
Bipolar	A	11	2	55	32	43	83	42	83	7	7.4	0.892	n.s.
psychosis	B	16	3	51	30	46	82	37	76	10	10.2	1.181	n.s.
(n = 89)	C	17	0	49	34	51	100	41	75	0	18.7	2.195	0.01
	D	37	5	29	29	66	89	50	44	13	36.1	3.597	0.001
	E	43	7	23	27	70	86	53	*36*	*20*	*39.5*	3.744	0.001
	F	44	9	22	25	69	83	52	34	27	35.9	3.325	0.001
Schizo-	A	4	0	56	40	44	100	32	93	0	5.9	0.833	n.s.
affective	B	8	1	52	39	47	92	37	87	2	9.5	1.318	n.s.
psychosis	C	8	3	52	37	45	69	41	87	9	3.6	0.496	n.s.
(n = 139)	*D*	26	8	34	32	58	75	48	*57*	*21*	*19.9*	2.494	0.05
	E	32	14	27	27	59	70	49	46	34	19.2	2.301	0.05
	F	39	15	25	25	60	70	50	42	38	19.5	2.306	0.05

TP = True positive; FP = false positive; FN = false negative; TN = true negative.

$1 = \frac{a+d}{a+b+c+d}$ = 'overall efficiency'; $2 = \frac{a}{a+b}$ = 'specificity'; $3 = \frac{d}{c+d}$ = 'sensitivity'; $4 = \frac{c}{a+c}$ = 'false negativity'; $5 = \frac{b}{b+d}$ = 'false positivity'.

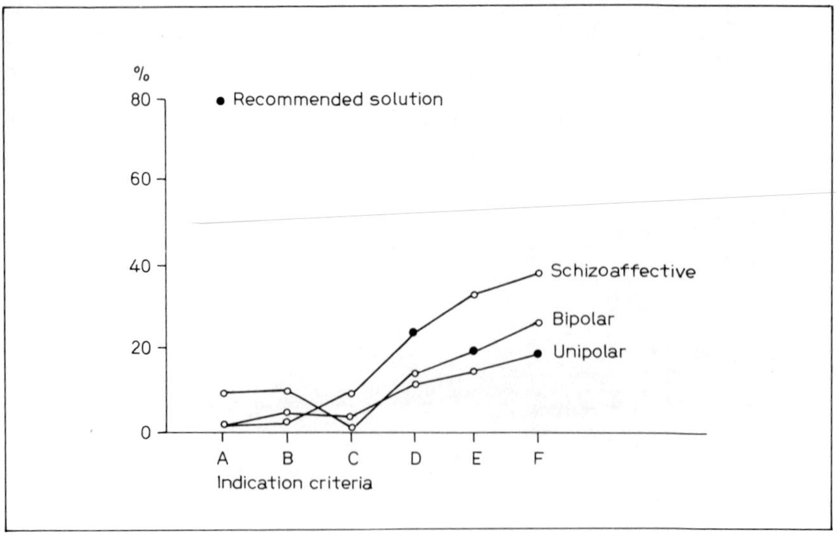

Fig. 3. Overtreatment = 'false positivity' = $\frac{\text{'false positives'}}{\text{'all negatives'}} = \left(\frac{b}{b+d}\right)$.

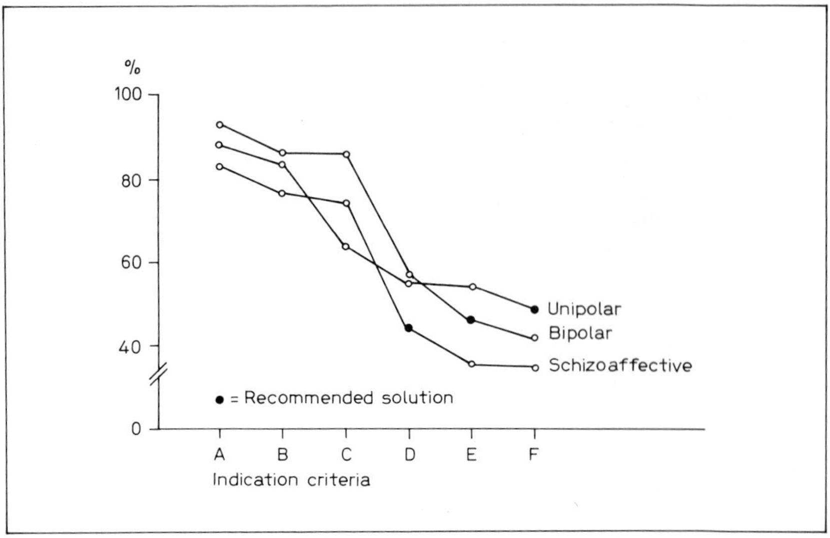

Fig. 4. Undertreatment = 'false negativity' = $\frac{\text{'false negatives'}}{\text{'all positives'}} = \left(\frac{e}{a+c}\right)$.

Table VI. Recommendations for selecting patients for lithium prophylaxis

Besides the index episode the patient must have experienced at least one earlier episode within a defined period of time, i.e.:

Unipolar depression	5 years
Bipolar manic-depressive psychoses	4 years
Schizoaffective psychoses	3 years

Table VII. Reduced patient sample

	Without lithium	With lithium	Total number
Unipolar	157	2	159
Bipolar	85	10	95
Schizoaffective	22	128	150

Duration of Lithium Prophylaxis

In practice, we are not only dealing with the question of which patient should be treated with lithium, but also how long should prophylactic lithium treatment be maintained? Of course, in many cases the prophylactic effect of lithium is incomplete, i.e. despite the long-term medication the patient suffers from further episodes which, however, may be attenuated in intensity and duration, and the free intervals may have become longer. It is obvious that in these cases prophylaxis has to be continued.

In other cases where the patient has become symptom-free, the difficult question arises as to whether and how lithium should or could be stopped. This is linked with the question of whether psychotic activity decreases with age or not. Some of our statistical data may be helpful in this difficult decision.

Our questions are: How active is the long-term natural course of affective disorders, especially in the second half of life? Is there a decrease or increase of the relapse risk? Is there an age limit where we can assume that lithium is no longer required? Our investigation is based on patients who 'never' received lithium as prophylactic treatment. (All patients with lithium treatment of more than 6 months' duration were excluded.) Table VII shows the number of patients who were included and excluded from our study. The schizoaffective group was also excluded from this analysis. Their course is very similar to the bipolar group, and the data have not yet been analyzed.

Methodology

Analyzing our lithium-free sample of 157 unipolar and 85 bipolar patients, we wanted to take into account, to a certain extent, the influence of age and age at onset. First of all we plotted all the episodes into age classes of 5 years. From this data we can derive three criteria of morbidity: (1) the percentage of patients with relapses in the sense of presence of episodes in each 5-year period according to age; (2) the frequency of episodes per year for each 5-year period according to class, and (3) the duration of the episodes as a percentage of the observation for each 5-year class: this analysis has not yet been carried out.

In order to take into account the influence of the age at onset in a crude way, we distinguish between early onset patients, i.e. age at onset below 40, and late onset patients, i.e. age at onset of 40 years or more.

Results

Percentage of Relapsing Patients

Figure 5 shows the percentage of unipolar patients with relapses calculated for each 5-year period. In order to exclude a bias in each case, the first 5 years including onset of the disorder were excluded. Thus, only relapses after the first 5 years were registered. Until the age of 64, about 50% of early onset unipolar patients suffer from relapses. Afterwards the relapse risk decreases to about 25% of the sample, but our figures in this subgroup are very small. Among late onset unipolar patients there is a relapse risk between 40 and 45% within the age period of 50 and 74 years showing a decrease after the age of 75 years to about 25%.

Figure 6 shows the corresponding data for bipolar patients broken down into early and late onset cases. Surprisingly, the late onset bipolar patients show a higher occurrence of relapses, but 70–80% of the patients relapse every 5 years until the age of 74 years. Later there may be a decrease to about 30%. The early onset group shows a relapse rate of about 50% of the sample until the age of 64 years, afterwards we find a decrease to about 25%.

If using these curves we try to reach a conclusion for lithium prophylaxis, we may state that: early onset unipolar and bipolar cases have to be treated until the age of 65 and late onset cases until the age of 75 years.

Frequency of Episodes per Year

More detailed information about the relapse risk can be calculated based on the frequency of observed episodes per year. The average frequencies for each decade are given in figure 7. Among unipolar patients the average relapse rate is exactly 0.24 per year, it means that on average one episode has been observed every fourth year. This is true for both subgroups, early and late onset patients, until the age of 69 years. Among the late onset cases there may be a later decrease of the risk. But this conclusion is not well founded because the length of episode may be longer at this age and reduce the number of new episodes observed per year.

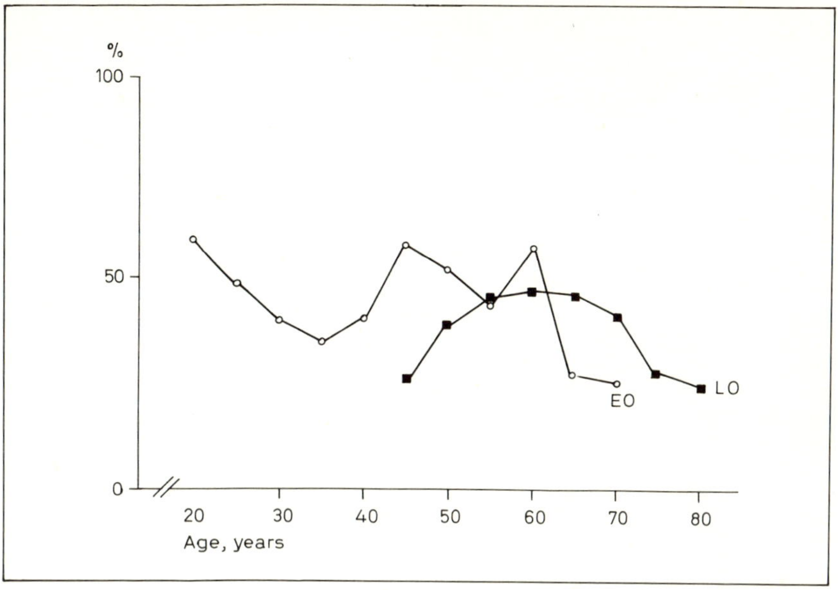

Fig. 5. Percentage of relapsing unipolar patients early onset (EO) versus late onset (LO).

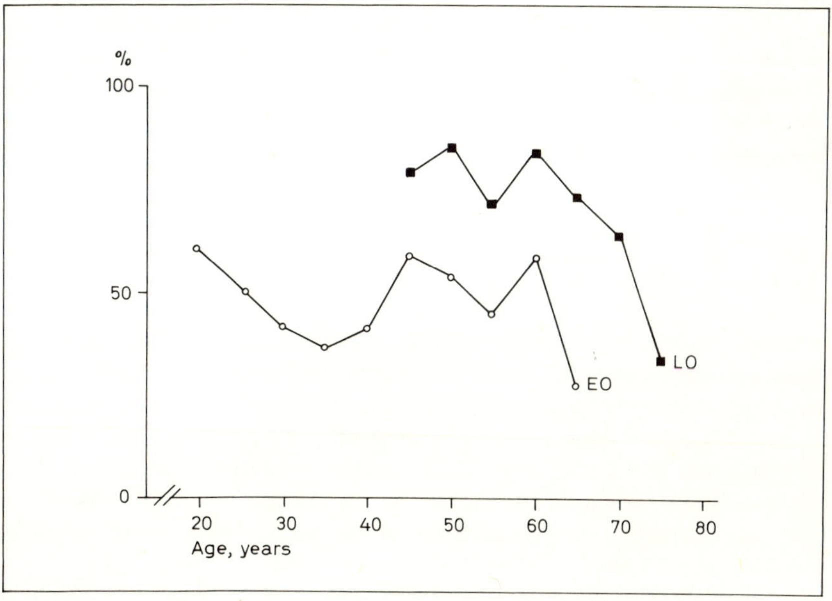

Fig. 6. Percentage of relapsing bipolar patients. Early onset (EO) versus late onset (LO).

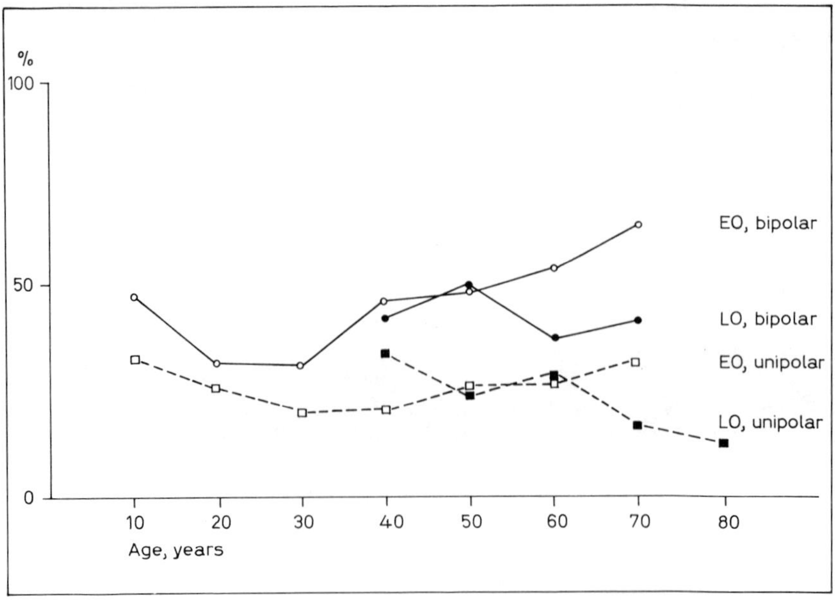

Fig. 7. Episodes per year in unipolar and bipolar patients. Early onset versus late onset.

Among bipolar patients we found on average 0.51 episodes per year which means that one episode was observed every second year. Until the age of 74 years we could not find a decrease of the risk. The relapse risk of the early onset group seems even to increase with age. The conclusion for lithium treatment is again that from a statistical point of view it has to be maintained until the age of 75 years.

Conclusions

This statistical analysis treats two problems: (1) Which patients suffering from affective psychoses should be treated prophylactically with lithium? (2) How long should the treatment be maintained?

A prospective study on the course of affective disorders was carried out in 159 unipolar, 95 bipolar, and 150 schizoaffective patients. Assuming that lithium prophylaxis is necessary if a patient has to expect a relapse risk of at least two episodes within 5 years after the initial episode we come to the following conclusions: (1) About 40% of the unipolar depressive patients, and $^2/_3$ of the bipolar manic-depressive, and schizoaffective patients require lithium prophylaxis. (2) Besides the initial episode the patient must have experienced at least one earlier episode within a defined period of time: i.e. unipolar depression, 5 years; bipolar manic-depressive psychoses, 4 years; and schizoaffective psychoses, 3 years.

Thus, a patient must have suffered from only one further episode besides the initial episode within the recommended period of time in order to fulfil the prerequisites. Applying these instructions an undertreatment of 36–57% of the patients who really require treatment still occurs, and there is an overtreatment in 19–21% of the patients who receive lithium and do not really need it.

In a second approach the long-term psychotic activity was analyzed. The proportion of relapsing patients and relapses (episodes) per year were computed taking into account age and age at onset. Until the age of 75 no decrease of psychotic activity was found in unipolar and bipolar disorders. Therefore, increasing age alone does not justify stopping prophylactic lithium treatment.

Concluding this paper, we must say that further investigation in this field is highly desirable.

Reference

Angst, J.: Verlauf unipolar depressiver, bipolar manisch-depressiver und schizo-affektiver Erkrankungen und Psychosen. Ergebnisse einer prospektiven Studie. Fschr. Neurol. Psychiat. *48:* 3–30 (1980).

J. Angst, MD, Research Department, Psychiatric University Clinic,
CH–8008 Zürich (Switzerland)

Adv. biol. Psychiat., vol. 7, pp. 230–239 (Karger, Basel 1981)

Antidepressants and Psychotherapy in Depression[1]

Myrna M. Weissman

Department of Psychiatry and Epidemiology, Yale University School of Medicine, and Depression Research Unit, Connecticut Mental Health Center, New Haven, Conn., USA

I appreciate that a society of biological psychiatrists can be sufficiently broad to include a paper on the treatment of depression with psychotherapy. I also appreciate that, even though you are skillful and humane therapists, you have at least two ideologic biases about psychotherapy.

First, there is a theoretical bias. You are skeptical about the relevance of psychosocial interventions in a disorder which your own research and that of others increasingly have demonstrated genetic and biochemical components.

Second, there is an empirical bias. Whereas you are aware of the overwhelming evidence for the efficacy of pharmacotherapy for depression, particularly the tricyclics, you are skeptical or unaware of the growing (albeit small) body of evidence for the value of psychotherapy in depression.

In this paper I won't deal with the first issue – the interplay between psychosocial and biological genetic factors in a complex disorder such as depression. This interplay might be a discussion for a future session of this group that could focus on the parallels between depression and other chronic, noninfectious medical disorders such as hypertension, heart disease, or arthritis. A useful epidemiologic concept which would highlight the parallels between these disorders would be that of relative risk, i.e., the relative contribution of various risk factors (biologic and psychosocial) which increase the probability of the onset or the exacerbation of the disorder.

Instead, I'd like to focus on the empirical evidence for psychosocial interventions – the growing body of scientific evidence for the efficacy of psychotherapy alone, in comparison, and in combination with antidepressants in the treatment of ambulatory depression.

In the scientific community, the interest in this topic is not limited to psychiatric disorders such as depression. Recently, the Institute of Medicine of the National Academy of Sciences in Washington, D.C., sponsored a conference

[1] This research was supported in part by Alcohol, Drug Abuse, and Mental Health Administration grant MH 26466 from the Clinical Research Branch, National Institute of Mental Health, Rockville, Md., USA.

on the psychosocial and pharmacologic interventions which included the treatment of depression, hypertension, and diabetes.

Recent Developments in Testing of the Psychotherapies

This discussion could not have taken place if not for the acceleration in the development and scientific testing of the psychotherapies over the last decade. These developments, which are being used in the testing of psychotherapy, include:

(1) clinical trials with random assignment of patients to treatment; (2) well-defined protocols with inclusion and exclusion criteria; specification of the amount, timing, and type of treatment, criteria for withdrawal of patients from treatment; (3) incorporation of the new advances in psychopathology including more reliable diagnostic measures; (4) testing of therapies in diagnostically homogeneous patients in order to determine whether a particular therapy is useful for a particular disorder; (5) recognition and use of various control groups to compare against psychotherapy; (6) evaluation of outcome by persons who are independent of and blind to the treatment condition; (7) use of videotapes of psychotherapy sessions to independently assess the quality of the treatment and the outcome; (8) use of standardized, quantified measures which cover a variety of dimensions of outcome, including symptoms and social functioning; (9) evaluation of the possible negative side effects of psychotherapy; (10) specification of the psychotherapies in procedural manuals which record the therapeutic techniques and their sequence, in order to insure that a uniform treatment and the one under study is being used consistently in the research, both within and between therapists, and (11) development of training programs for therapists involved in clinical trials to insure that the procedures specified in the manual are being transmitted.

As a result of these efforts, evidence for the efficacy and safety of psychotherapy and pharmacotherapy in comparison and in combination has been gradually accumulating. Increasingly sophisticated questions are now being answered about what psychotherapy, in combination with what tricyclic, in what sequence, for what effect, and for which subtype of depressed patient.

Most of the clinical trials of psychotherapy have been conducted with ambulatory depressed patients who are not psychotic and do not experience manic episodes, i.e., the nonbipolar depressed patient. This emphasis derives from the high prevalence and associated moribidity of ambulatory depression, and the economics of reimbursement (the United States Senate Finance Committee has begun to lean on the profession to provide proof of the efficacy of a variety of treatments, including psychotherapy, before expanding reimbursement for them).

Against this background, I'd like to describe the psychotherapies which have been designed and tested with the ambulatory depressed patient.

Description of the Psychotherapies

Although many types of psychotherapies are widely and appropriately used in depression, as of this writing, five therapies (cognitive, interpersonal, behavioral, marital, and group) have undergone scientific testing in clinical trials using homogeneous samples of depressed patients who meet the criteria of major depression as defined by the research diagnostic criteria (RDC) [1] and the diagnostic and statistical manual, 3rd ed. (DSM-III).[2]

Cognitive Therapy

Cognitive therapy, developed by *Beck* [2], is based on the assumption that the affective response in depression is determined by the way an individual perceives experience. As a result of an emergence of maladaptive cognitive themes, the depressed patient tends to regard himself and his future negatively. Correction of negative concepts is expected to alleviate the depressive symptoms. For example, an extremely low self-concept can be treated by presenting a hierarchy of cognitive tasks, and through these tasks, demonstrating the invalidity of the patient's self-reproaches.

Interpersonal Psychotherapy

Interpersonal psychotherapy assumes that the development of depression occurs in a social and interpersonal context and is determined by the interpersonal relations between the depressed patient and significant other [3]. Depression is seen as having three component processes: symptom formation, social adjustment, and personality. Interpersonal psychotherapy attempts to intervene in the first two, and does not claim to have an impact on enduring aspects of personality.

The goals are to improve the quality of the patient's social and interpersonal functioning by enhancing ability to cope with internally and externally induced stresses, by restoring morale, and by helping the patient to deal with the personal and social consequences of the disorder. Thus far, techniques have been developed for the management of depressed patients with grief and loss, interpersonal deficits, interpersonal role disputes, usually marital disputes, and role transitions.

[2] The RDC for the syndrome of major depression includes dysphoric mood, at least five other symptoms, persistence of at least 2 weeks, producing impairment of functioning or necessity of treatment, and occurring in the absence of schizophrenia or other disorders which may better explain the symptoms [1].

Behavioral Approaches

Behavior therapy explains the occurrence of depression in stimulus and response terms. A low rate of positive reinforcement elicits certain aspects of the depressive syndrome and is predicted by negative reinforcing events in the environment and lack of available positive reinforcements as well as by the individual's behavior [4].

Several techniques based on behavioral concepts have been developed for depressed patients. These are: social skills therapy, which emphasizes increasing assertiveness, verbal skills, and social adjustment; pleasant events, which focus on increasing pleasant and rewarding experiences; and self-control, which emphasizes self-monitoring, self-evaluation, and self-reinforcement to correct the depressed patients's problem in self-control in monitoring negative events, and in making internal attributions of responsibility and excessive self-punishment.

Marital Therapy

Marital therapy attempts to alter the interaction between or among the marital partners. It is based on the assumption that the marital relation affects thoughts, feelings, and behavior, and that symptom removal can be achieved by changing the transactional marital system. Marital therapy can be individual, collaborative, concurrent, conjoint, combined or group. The theoretical base and a typology of marital types have been well summarized by *Gurman and Kniskern* [5].

Group Therapy

In group therapy a trained psychotherapist and a group of patients attempt to effect changes in the emotional states, attitudes, and behavior of the patient. Currently, explicit definitions of techniques and descriptions of what occurs within the group format are lacking, although there are efforts to develop cognitive therapy in a group context.

Status of Development of the Five Psychotherapies

Table I shows the status of the development and testing of these five therapies [6]. Cognitive therapy has been tested in six trials; interpersonal psychotherapy in three trials; behavioral approaches in eight trials; marital and group therapy in one trial each. Of the five therapies, only cognitive, interpersonal, and some of the behavioral techniques have been specifically designed for depressed patients, and procedural manuals are available only for the cognitive and interpersonal therapies. However, this is a rapidly moving field. Some of the trials noted are still in progress. Undoubtedly, other clinical trials, new therapies, and more procedural manuals are under development or being completed, as of this writing.

Table I. Status of development of the five psychotherapies

	Cognitive	Interpersonal	Behavioral Approaches	Marital	Group
Number of trials with depressed patients	6	3	8	1	1
Designed for depressed patients	yes	yes	(some)	no	no
Procedural manual	yes	yes	no	no	no

Table II. Evidence for efficacy of psychotherapy and drugs in the treatment of depression

	Number of studies
Psychotherapy alone	
Psychotherapy > control	9
Psychotherapy = control	0
Psychotherapy < control	0
Psychotherapy in comparison to drugs	
Psychotherapy > drugs	1
Psychotherapy = drugs	1
Psychotherapy ? drugs	3
Psychotherapy > drugs in social functioning	
Psychotherapy < drugs in symptom prevention	
Psychotherapy in combination with drugs	
Psychotherapy + drugs > psychotherapy > drugs > control	4

Evidence for Efficacy

Table II summarizes the evidence for efficacy of the five psychotherapies alone, in comparison, and in combination with drugs for the ambulatory depressed patient. A more detailed description of these results can be found in *Weissman* [6].

Psychotherapy Alone

There are nine studies testing the efficacy of psychotherapy in comparison with a low-contact or nonactive treatment control group. All five treatment approaches are represented. All of the studies support the efficacy of psychotherapy alone as compared with a control group. In none of the studies was psychotherapy equal or worse than a control or a nontreatment group.

Psychotherapy in Comparison to Drugs

There are five completed studies from which information on psychotherapy in comparison with a tricyclic antidepressant can be drawn. All five of the psychotherapies are represented. One study found psychotherapy (cognitive therapy) superior to drugs (imipramine) in attrition and symptom reduction for acute treatment of depressed patients. One study found drugs and psychotherapy about equal for acute symptom reduction. Three studies found drugs superior to psychotherapy (interpersonal, group or marital) in the prevention of relapse or symptom reduction, but psychotherapy slightly superior to drugs in the enhancement of social functioning. The psychotherapy effect in one study occurred only in patients who remained in treatment for 8 months without relapsing.

The results from the comparisons of psychotherapy with drugs are equivocal for acute treatment. The treatments appear to be equal overall but to have different targets of action. The superiority of cognitive therapy to drugs in one study requires replication before the results can be accepted.

Psychotherapy in Combination with Drugs

There have been four completed studies testing the efficacy of psychotherapy (interpersonal, marital or group) combined with a tricyclic antidepressant. All four studies show the superiority of combined treatment over a control group or over either treatment alone. In none of the studies were there negative interactions in combining drugs with psychotherapy. The effects of the two treatments together were additive.

Studies in general have found overall that the combination of psychotherapy and a tricyclic is better than either treatment alone or than no regular treatment. The combination provides a broader spectrum of action as each treatment may affect different domains [7]. Psychotherapy seems to have its effect on social and interpersonal areas and has a slower onset of action. Drugs have their effect on the vegetative symptoms of depression such as sleep and appetite, and have a more rapid onset of action.

Implications for the Management of Depression

The overall evidence thus far is in favor of the combined treatment (drugs and psychotherapy) over either treatment alone for the ambulatory nonbipolar depressive. However, there is considerable variability in patients' acceptance and preference for the various treatments and there are different types of depression. Some patients want only one of the treatments. They may be reluctant to take medication or may not wish to talk about personal matters. Other patients may be able to tolerate only one of the treatments regardless of preference. They may be intolerant of the medication's side effects or have cognitive deficits or economic problems which make psychotherapy impractical. What research evidence can guide the management of the ambulatory depressive with drugs and psychotherapy?

Which Psychotherapy?

We cannot yet state how the psychotherapies compare separately or to one another, nor can we offer guides on the basis of research data about the selection of a particular psychotherapy for the depressed patient. Such data will be forthcoming in the next 3–5 years.

Pharmacotherapy versus Psychotherapy

The evidence comparing pharmacotherapy to psychotherapy is equivocal and inconclusive. However, the patient who will not accept one treatment should not be denied the other, since the evidence shows that either treatment is better than no treatment. Moreover, there are several recent findings on the symptoms, social functioning, and diagnostic subtypes of patients from our own data and that of others which can begin to guide the clinician's selection of an effective treatment.

The research evidence shows that pharmacotherapy is most efficacious on sleep disturbance and appetite loss, and that the effects on these symptoms are relatively early (1–4 weeks). Therefore, the patient whose depression is mostly manifest in these symptoms should not be denied an antidepressant [7].

Alternately, psychotherapy is most efficacious on suicidal feelings, guilt, loss of interest, and social and interpersonal problems, and is indicated for patients whose depression is mostly manifest in these areas. Psychotherapy has a later onset of action and the differential effects on social and interpersonal functioning may not be apparent for 6 months to 1 year.

Recent findings by *Prusoff* et al. [8] from the New Haven-Boston collaborative study suggest that the RDC subtypes of endogenous and situational depression may be useful guides to treatment selection.

Endogenous and situational depressions are defined by the RDC [1] as follows:

Endogenous Major Depressive Disorder. From groups A and B a total of at least four symptoms for probable, six for definite, with at least one symptom from group A.

Group A. (1) Distinct quality to depressed mood, i.e., depressed mood is perceived as distinctly different from the kind of feeling the patient would have or has had following the death of a loved one; (2) lack of reactivity to environmental changes (once depressed the patient does not feel better, even temporarily, when something good happens); (3) mood is regularly worse in the morning; (4) pervasive loss of interest and/or pleasure.

Group B. (1) Feelings of self-doubt or reproach or excessive or inappropriate guilt; (2) early morning awakening or middle of night insomnia; (3) psychomotor retardation or agitation (more than a mere subjective feeling of being slowed down or restless); (4) poor appetite; (5) weight loss (2 pounds a week over several weeks or 20 pounds in a year when not dieting); (6) loss of interest or pleasure (may not be pervasive) in usual activities or decreased sexual drive.

Situational Major Depressive Disorder. Illness has developed after an event or in a situation that seems likely to have contributed to the appearance of the episode at that time. In making the judgment the amount of stress inherent in the event or situation is considered, the cumulative effect of such stresses, and the closeness of the events to the onset or exacerbation of the depressive episode.

In the New Haven-Boston clinical trial, 81 depressed patients were randomized into one of four treatments: combined amitriptyline and interpersonal psychotherapy, either treatment alone, or a nonscheduled treatment control group [7]. Of the 81 patients, 31 (40%) were diagnosed by the psychiatrist at the intake interview as situational only, and 20 (27%) as endogenous only.

Figure 1 shows the results on overall depressive symptoms based on the Hamilton depression scale at the end of treatment for the depressed patients diagnosed as situational only. These patients responded equally well to all treatments and responded poorly to nonscheduled treatment. However, the combination of drugs and psychotherapy did not seem to offer more than psychotherapy alone.

Figure 2 shows the results for the endogenous depression. As can be seen, these patients responded poorly to psychotherapy alone which was not any better than no scheduled treatment. The response was best to the combination of drugs and psychotherapy. The lack of better response to drugs alone may have been due to poor compliance. This study was begun in 1972 before plasma level monitoring of tricyclics was feasible. Future studies will incorporate plasma level tests.

These results are based on small samples and require replication. They suggest that the situational depressions respond best to psychotherapy alone and the addition of drugs adds little. The endogenous depression, on the other hand, does not do well on psychotherapy alone.

Future Directions

Despite the progress, there are gaps in the data. There are many psychotherapies which, on the basis of clinical experience, we would expect to be quite efficacious in the treatment of depression, but which have not as yet been tested in clinical trials. We also do not know the value of the treatments described for the bipolar patient or for the psychotically depressed or hospitalized patient.

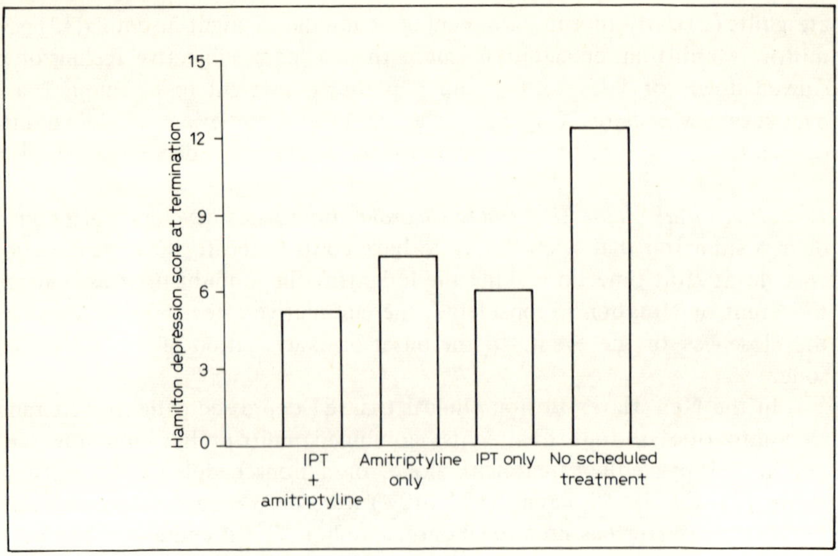

Fig. 1. Situational depression only. IPT = Interpersonal psychotherapy.

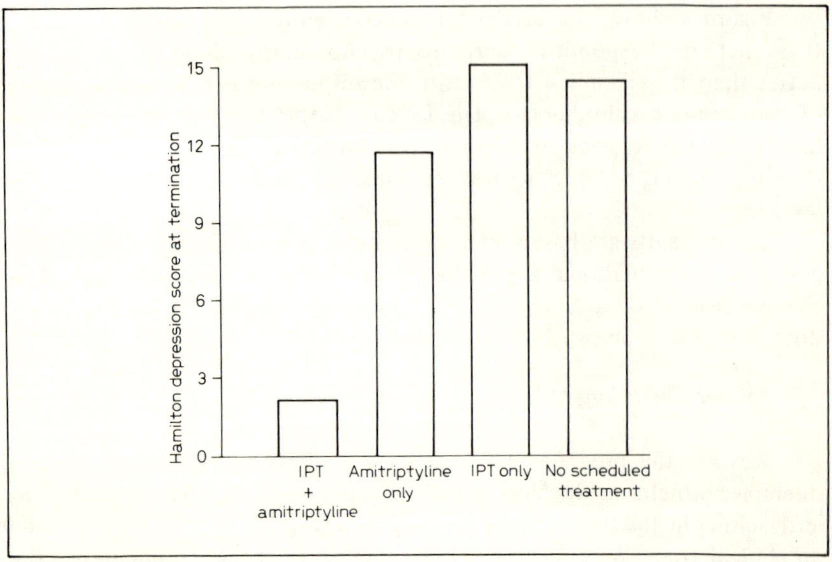

Fig. 2. Endogenous depression only.

Although one study, while not random assignment, did suggest the value of couples therapy in conjunction with lithium in the management of married bipolar patients [9]. Those couples in group therapy had a more benign posthospitalization course than those given minimum support.

There is, however, sufficient evidence from small individually conducted clinical trials to launch a multi-center collaborative trial. Such a study has been designed by the staff at the Clinical Research Branch of the National Institute of Mental Health, in consultation with many experts, and is currently under way. This collaborative study involves three clinical sites and incorporates all of the recent scientific advances described for the conduct of clinical trials. In addition, a number of single-center clinical trials are currently under way. In the next 3 years data will become available from at least five studies in the United States. The next decade will bring clearer answers about which types of psychotherapies to use with which drugs, and for which depressed patients.

References

1 Spitzer, R.L.; Endicott, J.; Robins, E.: Research diagnostic criteria: rationale and reliability. Archs gen. Psychiat. *35:* 773–782 (1978).

2 Beck, A.: Cognitive therapy and the emotional disorders (International Universities Press, New York 1976).

3 Klerman, G.L.; Rounsaville, B.; Chevron, E.; et al.: Manual for Short Term Interpersonal Psychotherapy (IPT) of Depression, 4th Rev., 1979 (unpublished).

4 Lewinsohn, P.; Biglan, A.; Zeiss, A.: Behavioral treatment of depression; in Davidson, The behavioral management of anxiety, depression and pain, pp. 91–146. (Brunner/Mazel, New York 1976).

5 Gurman, A.S.; Kniskern, D.P.: Research on marital and family therapy: progress, perspective and prospect; in Garfield Bergin, Handbook of psychotherapy and behavioral change: an empirical analysis; 2nd ed. (Wiley & Sons, New York 1979).

6 Weissman, M.M.: Psychological treatment of depression: evidence for the efficacy of psychotherapy alone, in comparison with and in combination with pharmacotherapy. Archs gen. Psychiat. *36:* 1261–1269 (1979).

7 DiMascio, A.; Weissman, M.M.; Prusoff, B.A., et al.: Differential symptom reduction by drugs and psychotherapy in acute depression. Archs gen. psychiatry *36:* 1450–1456 (1979).

8 Prusoff, B.A.; Weissman, M.M.; Klerman, G.L., et al.: Research diagnostic criteria: their role as predictors of differential response to psychotherapy and drug treatment. Archs gen. Psychiat. *37:* 796–801 (1980).

9 Davenport, Y.B.; Ebert, M.H.; Adland, M.L.; Goodwin, F.K.: Couples group therapy as an adjunct to lithium maintenance of the manic patient. Am. J. Orthopsychiat. *47:* 495–502 (1977).

M.M. Weissman, PhD, Director, Depression Research Unit,
Connecticut Mental Health Center,
904 Howard Avenue, Suite 2A, New Haven, CT 06519 (USA)

Subject Index